AF539424

APPLICATIONS OF SPECTROSCOPY

APPLICATIONS OF SPECTROSCOPY

By

Dr. K. Sonamuthu

Reader
Deptt. of Physics
J.N.R. Mahavidyalaya
Port Blair
Andaman and Nicobar Islands

DISCOVERY PUBLISHING HOUSE PVT. LTD.
NEW DELHI-110 002

First Published – 2010

Reprinted – 2024

ISBN: 978-81-8356-604-9

© Author

Applications of Spectroscopy

Published by:

DISCOVERY PUBLISHING HOUSE PVT. LTD.

4383/4B, Ansari Road, Darya Ganj
New Delhi-110 002 (India)
Phone: +91-11-23279245; 23253475; 43596065
Mobile: +91 9811179893 / +91 9871656464
E-mail: discoverybooksindia@gmail.com
discoverypublishinghouse@gmail.com
orderdphbooks@gmail.com
web: www.discoverypublishinggroup.com

Printed at:
Infinity Imaging Systems
Delhi

Preface

It is generally believed, that is needed to signpost the way to the development of materials with high critical temperatures, even to room temperature and above, is the formulation of a proper theoretical understanding of superconductivity in theses materials. Many experimental studies suggest an intimate relationship between phonons, and the superconducting transition in the high T_c oxides. For example, Raman frequencies and line widths, elastic constants and thermal conductivity all show changes around T_c. Nevertheless, there is not a well-developed theory to explain how the phonons are related to T_c. In conventional superconductors, the role of phonons is well established. However, in high T_c superconductors, the role played by phonons is controversial. Even if they do not mediate the pairing mechanism, they are likely to play an important subsidiary role. Hence there is much interest in the phonon structure of these materials. In this work, an attempt has been made to study the different rare-earth substituted systems using normal coordinate analysis methods.

The normal state and superconducting properties are believed to arise form the strongly correlated motion of the electronic charge carriers (electrons or holes) in the CuO_2 planes that essentially define the layered Cuprate HTSC. The other cations and oxygen atom in the structure provides structural stability and control the number of charge carriers in the CuO_2 planes. Literally, hundreds of novel superconducting CuO_2 planar compounds have already been synthesized, most of which are substitutional variants of about a dozen basic chemical structure.

The small pinning energy means that flux lines can be relatively easily unpinned by intrinsic thermal fluctuations. This is the main challenge for high-temperature, high current power applications of HTSC. The Raman spectroscopy of HTSC explains that the role played by the phonons in the mechanism of superconductivity in the perovskite family of superconductors is not fully understood although a large series of experimental results has shown considerable coupling of some phonons to electronic excitation in these systems. Optical spectroscopy and especially Raman and infrared spectroscopy revealed strong direct influences of the changes in the electronic states to the phonon at the center of the Brillouin zone.

It is concluded that the observed frequencies in Raman and infrared spectra gives the support of the presence of symmetrical structures of high temperature superconductors. It also supports for the strong electron-phonon interaction in the high temperature superconductors are discussed in this book.

Dr. K. SONA MUTHU, Ph.D., D.R.T.E.

Contents

CHAPTER 1

Preview on High Temperature Superconductors

INTRODUCTION

The exotic phenomenon of superconductivity began with the discovery of Zero resistance by Kamerlingh Onnes in the year1911. Before 1986, superconductivity was a phenomenon restricted to metals, occurring only at very low temperature and requiring liquid helium as a refrigerant. The highest known critical temperature, T_c, was 23K for inter-metallic compound Nb_3Ge. In the early part of 1986, Bednor and Muller found that a mixed oxide of lanthanum, barium and copper became superconducting at a critical temperature of about 30K. This work, for which they received the Noble Prize for physics in1987, heralded the high T_c explosion. Later, Chu *et al.* showed that T_c of La-Sr–Cu-O could be raised from 35K to 50K by applying pressure, a rare earth ion, namely Yttrium was used, which is smaller than Lanthanum and thus a higher T_c was obtained owing to the included internal pressure. This resulted in the discovery of a new compound subsequently confirmed as $YBa_2Cu_3O_7$, with a T_c of 90K. The intense research effort thus unleashed has led to the discovery of further compounds: Bi-Sr-Ca-Cu-O with T_c up to 120K and Tl-Ba-Ca-Cu-O with T_c as high as 130K.

In March 1993, the mercury copper oxides $HgBa_2CuO_{4+x}$(Hg-1201) came to light, the latest of the board has a set a record of 133K for the superconducting transition temperature. Recently, Chu *et al.* have shown that the Mercury-based systems have a T_c of 160K and this is believed to be the highest critical temperature achieved so far. Although the majority of high temperature superconductors (HTSC) are P-type, involving hole doping of the planes, a number of N-type electron doped HTSC such as $Nd_{2-x}Ce_xCuO_4$ have also been discovered.

The critical current density is the most important parameter of the superconductor for the majority applications. In the case of a normal conductor, resistance arises because electron lose momentum which they have acquired from the electric field, by scattering collisions with defects of the crystalline lattice. In the superconducting state, these collisions still occur but pairs of electron act in a co-operative manner to ensure that there is no net change in momentum. In conventional superconductors, the motion of the two electrons in a pair is correlated by a lattice phonon. The behaviour of these pair is described by the Bardeen, Cooper and Schrieffer (BCS) theory of super superconductivity. BCS theory predicts an energy gap, k_bT_c of about 10^{-3} eV, in the band structure of a superconductor at the Fermi surface.

Some interesting systematic emerged from the studies of the bismuth and thallium compounds. In 1988, superconductivity was reported in oxide systems based on Bi, Cu and alkaline earth ions. These compounds do not contain rare-earth elements like the earlier ones. They belong to a structural family with the ideal formula $Bi_2Sr_2Ca_{n-1}Cu_nO_{4n+2+8}$ containing variable number, n, of copper ($Cu\text{-}O_2$) layers.

In the thallium system, there are two series of compounds, $Tl_1Ba_2Ca_{n-1}Cu_nO_{2n+3}$ and $Tl_2Ba_2Ca_{n-1}Cu_nO_{2n+4}$ (n=1 to 5), which can be modeled as an interleaving of rock salt and defective perovskite layers. For example, $TlBa_2CuO_5$ can be thought of as a layer of copper oxide sandwiched between two layers of barium oxide, which in turn sandwiched between two layers of thallium oxide. This compound, 1201, is not superconducting. A second copper layer can be inserted into the sandwich provided; it is separated from the first by calcium layer. This gives the 1212 compound, which has a T_c of approximately 90 K. A second calcium layer allows the insertion of a third copper layer giving the 1223 compound with T_c increased to 116 K. The corresponding series compounds with double thallium layers have somewhat higher values of T_c, for 2223, the highest T_c is at 128 K.

The increase in T_c as more copper layers are introduced into the structure, fuelled speculation that the route to room temperature superconductivity is to add more copper layers to the structure. These hopes were dashed when the 1245 and the 2234 compound were prepared and found to have T_c of less than 120 K and 102 K respectively. Thus, it would appear that T_c is maximized in the Tl_1 compounds at four layers, and in the Tl_2 compounds at three copper layers. But there is considerable difficulty in preparing the pure form of each structure. In all of these compounds, intergrowth are common in the form of alternating regions of the different variants.

It is generally believed, that is needed to signpost the way to the development of materials with high critical temperature, even to room temperature and above, is the formulation of a proper theoretical understanding of superconductivity in these materials. Many experimental studies suggest an intimate relationship between phonons, and the superconducting transition in the high T_c oxides. For example, Raman frequencies and line widths, elastic constants and thermal conductivity all show changes around T_c. Nevertheless, there is not a well-developed theory to explain how the phonons are related to T_c. In conventional superconductors, the role of phonons is well established. However, in high T_c superconductors, the role played by phonons is controversial. Even if they do not mediate the pairing mechanism, they are likely to play an important subsidiary role. Hence there is much interest in the phonon structure of these materials.

The normal state and superconducting properties are believed to arise from the strongly correlated motion of the electronic charge carries (electrons or holes) in the CuO_2 planes that essentially define the layered cuprate HTSC. The other cations and oxygen atom in the structure provides structural stability and control the number of charge carries in the CuO_2 planes.

The Raman spectroscopy of HTSC explains that the role played by the phonons in the mechanism of superconductivity in the perovskite family of superconductors is not fully understood although a large series of experimental results has shown considerable coupling of some phonons to electronic excitation in these systems. Optical spectroscopy and especially Raman and Infrared spectroscopy revealed strong direct influences of the changes in the electronic states to the phonon at the centre of the Brillouin.

The present work is planned in four chapters. Chapter 1 gives brief introduction of high temperature superconductor, aim of the book and chapterization.

Chapter 2 deals with the General instructions of high temperature superconductor, with an emphasis on the two properties of superconductors. One is that the temperature below which the resistance of the material vanishes is called transition temperature T_c. The second fundamental property of a superconductor is that its normal resistance may be restored if a magnetic field greater than a critical value B_c is applied to the specimen. It is zero at T_c and as the temperature is reduced it increases, following the parabolic law $B_c = B_0\{1-[T/ T_c]^2\}$. The magnitude of B to achieve this is the critical induction. Further, the basic structural aspects of high T_c cuprates are also explained in this chapter. The cuprate superconductors like La-M-Cu-O, Y-Ba-CuO, Bi-Sr-Ca-Cu-O and Tl-Ba-Ca-Cu-O are widely worked on groups. Holes are the majority current carries in these systems. Finally this chapter ends with the recent developments, applications and future prospects.

In Chapter 3, the essential of normal coordinate analysis is discussed, which explains the picture of the normal coordinate analysis of the spectral frequencies of the high temperature of $LaCuO_4$ and Pr_x H_{01-x} Ba_2 Cu_3O_7 are presented in this chapter.

This chapter further deal with the normal coordinate calculation was performed using the programs GMAT and FPERT given by Fuhrer *et al.* The evaluated Raman frequencies agree close to the available observed infrared and Raman frequencies giving further support to the present assignment. The results and discussion are given in detail in the respective subsections.

To check whether the chosen set of vibrational frequencies makes maximum contribution to the potential energy associated with the normal coordinate frequencies of the superconducting material, the potential energy distribution was calculated using the equation:

$$PED = \frac{(Fij\ L2\ ik)}{\lambda k}.$$

Chapter 4 deals with the comparative study of phonon frequencies obtained from experimental and theoretical frequencies of Normal Coordinate Analysis of high temperature superconductors La_2CuO_4 and $Pr_xH_{01-x}Ba_2Cu_3O_7$. It is interesting to note that the experimental and normal coordinate analysis yield same phonon frequency a bridge between Solid State Physics and Spectroscopy to understand the mechanism of superconductivity.

Finally it is concluded that the observed frequencies in Raman and infrared spectra gives the support of the presence of symmetrical structures of high temperature superconductors. It also supports for the strong electron-phonon interaction in the high temperature superconductors are discussed and confirmed by potential energy distribution in this work.

CHAPTER 2

Review of Literature

ORIGIN OF HIGH TEMPERATURE SUPERCONDUCTOR

A superconductor shows vanishing resistance below a characteristic temperature called the transition temperature T_c. Soon after Onnes discovery in 1991, other researchers found that lead become superconducting below 7.2 K and tin below 3.7 K. Until 1986, the higher superconducting transition temperature was 23.21 K in Nb_3Ge as found by Cavater. The new era of high temperature superconductors began in 1986,with the announcement of T_c about 30 K by Nobel Prize winners, George Bednorz and Alexander Muller in Lanthanum-Barium-copper oxide [1] [La-Ba-Cu-O systems]. But the actual break through occurred only after 76 years of the discovery of the phenomenon of superconductivity and has prompted unprecedented interest and excitement. This was by the discovery of superconductivity in an oxide with a T_c above the boiling point of liquid nitrogen by Paul Chu and his co-workers in 1987. Later, Chu *et al.*[2] showed that the T_c of La-Sr-Cu-O could be raised from 35 K to 50 K by applying pressure. Then as an attempt to mimic the effect of increased pressure by chemical substitution, a rare earth ion, namely Yttrium was used, which is smaller than Lanthanum and thus a higher T_c was obtained owing to the included internal pressure. This resulted in the discovery of a new compound, subsequently confirmed as $YBa_2Cu_3O_7$ with a T_c of 90 K.[3] The intense research effort thus unleashed has led to the discovery of further compounds: Bi-Sr-Ca-Cu-O with T_c up to 120 K [4] and Tl-Ba-Ca-Cu-O with T_c as high as 130 K[5], including the remarkable C_{60} based fullerene compounds [6] in which T_c in Rb-Tl-C has been claimed at 42.5 K. Then C_{60} based superconductors provide another totally new arena for studies of novel superconducting organic materials with great future promise for even higher T_c's. In March 1993, the mercury based layered copper oxides $HgBa_2CuO_{4+x}$ (Hg-1201) [7] came to light; the latest of the brood has set a record of 133 K for the superconducting transition temperature. Recently, Chu *et al.* [8] have shown that the mercury based systems have a T_c of 160 K and this is believed to be the highest critical temperature achieved so far. Although the majority of high temperature superconductors (HTSC) are p-type, involving hole doping of the planes, a number of n-type electron doped HTSC such as $Nd_{2-x}Ce_xCuO_4$ [9] have also been discovered.

BASIC CHARACTERISTIC OF SUPER-CONDUCTOR

A superconductor is characteristics by zero electrical resistance [10]. The temperature below with the resistance of the material vanishes is called transition temperature $T_{c.}$

The second fundamental property of a superconductor is that its normal resistance may be restored if a magnectic field greater than a critical value B_c is applied to the specimen. It is zero at T_c and as the temperature is reduced it increases, following the parabolic law

$B_c = B_0\{1-[T/T_c]^2\}$.

It tends to a constant value B_0 as T approaches 0 K.

Superconductors allow the flow of electric current under zero voltage. Beyond a certain critical current, the material develops, the voltage drop and original conduction is recorded. J_c is the largest current density the material can carry under zero voltage in the superconducting state. Associated with an electric current, there will be a magnetic field. Hence, if a superconductor carries a current such that the field that is produces is equal to B_c, the resistance of the sample is restored. So superconducting state is characterized by $J < J_c$ [11].

A superconductor exhibits perfect diamagnetism with B = 0 inside it. This phenomenon is called the meissener effect whereby a superconductor expels magnetic flux from within and becomes a perfect diamagnetic provided the external applied field is small. The conditions defining superconducting state are:

E = 0 (from the absence of resistivity)

B = 0 (from Meissner effect).

Depending upon the magnetic behaviour, superconductors are classified as Type I and Type II [12] (Fig. 2.1). The materials that always expel the flux completely before returning to the normal state are classified as Type I superconductors or pure superconductors. Here B_c, the strength of the applied magnetic field of the applied magnetic field is parabolic. B_0 is the extrapolated value of B_c at T = 0. Except for Vanadium and Niobium all elements are Type I superconductors.

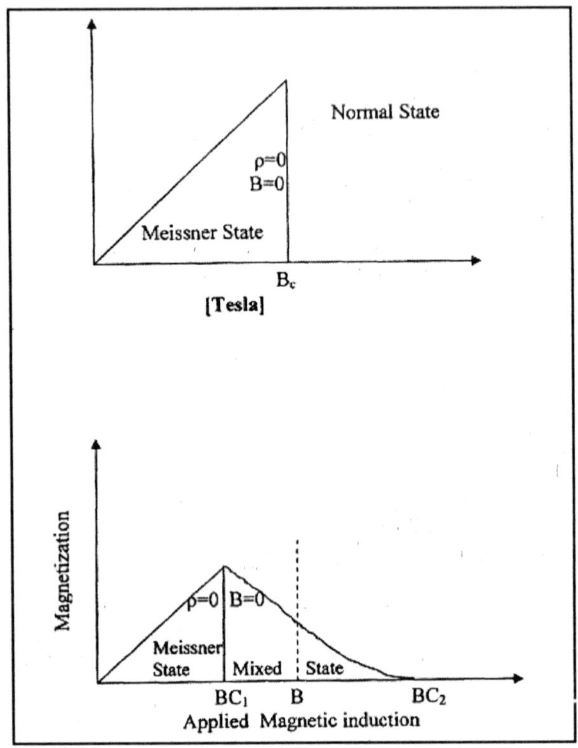

Fig. 2.1: Type I and II Superconductors

Type II superconductors remain superconduct-ing up to a very high magnetic field even with the penetration of the magnetic flux. There are two critical fields in the Type II superconductors: The lower critical field B_{c1} and the upper critical field B_{c2}. The flux is completely expelled only up to the field B_{c1}. So, in applied fields smaller than B_{c1}, Type II superconductors behave just like Type I superconductors. Above B_{c1}, the flux partially penetrates into the material until upper critical field B_{c2} is reached. Above B_{c2} the material returns to its normal state. The superconductor is in a mixed or vortex state between B_{c1} and B_{c2}. The Meissner effect is only partial. For all applied fields $B_{c1} < B < B_{c2}$, magnetic flux partially penetrates the superconducting specimen in the microscopic filaments called vortices.

PHENOMENON OF SUPERCONDUCTIVITY

Superconductivity, the complete loss of electrical resistance, is destroyed by the application of magnetic induction, B. The magnitude of B necessary to achieve this is the critical induction, B_c, and is a function of temperature rising from zero at T_c to a value $B_c(0)$ at T = 0 K. In the so-called type I superconductors, the magnetic flux is completely expelled from the interior of the superconductor if B is greater than B_c. This is called the Meissner effect. The passage of an electrical current greater than a critical value I_c also destroys superconductivity, I_c is just that current which results in the critical induction occurring at the surface of the superconductor and is therefore a function of temperature, applied magnetic field and conductor size. Critical inductions and their practical utility are limited. However, many alloy and compound superconductors have much higher critical inductions, in the range from 1-40 Tesla. These materials are termed as Type II superconductors and they have an incomplete Meissner effect. Magnetic flux enters the body of the superconductor in the form of localized flux lines, ach carrying a single quantum of magnetic flux, Φ_0. when a current, of density J flows in a Type II superconductors in a magnetic induction B, the flux line experiences a Lorentz force $F_L = J \times B$. The critical Lorentz force is equal to the total pinning force per unit volume F_p, which is a function of both temperature and magnetic induction.

The critical current density is the most important parameter of the superconductor for the majority of applications. In the case of a normal conductor, resistance arises because electron loses momentum which they have acquired from the electric field, by scattering collisions with defects of the crystalline lattice. In the superconducting state, these collision still occur but pairs of electrons act in a co-operative manner to ensure that there is no net change in momentum. In conventional superconductors, the motion of the two electrons in a pair is described by the Bardeen, Cooper and Schrieffer (BCS) theory of superconductivity.

BCS Theory of Superconductivity

The microscopic theory put forward by Bardeen, Cooper and Schriefer (BCS), in 1957, provides the better quantum explanation of superconductivity and accounts very well for all the properties exhibited by the superconductors. This theory involves the electron interactions through phonons as mediators. The basis of the formulation of BCS theory are the two experimental conclusion namely the isotope effect and the variation of specific heat of superconductor. This all suggest that non-zero transition temperature is a consequence of the finite mass of the ions which can contribute phonons by their vibrations. One can interpret the interaction between the lattice and electron as the constant emission and re-absorption (creation and annihilation) of phonons. BCS showed that the basic interaction responsible for superconductivity appear to be that of a pair of electrons by means of an interchange of virtual phonons. The lowering of electron energy implies that the force between the two electrons is attractive. This type of interaction is strongest when the two electrons have equal and opposite momenta and spins. This interaction can also be interpreted as the electron-electron interaction through phonons as the mediator. If the phonon energy exceeds electronic energy, the interaction is attractive.

The fundamental postulate of BCS theory is that superconductivity when occurs an attractive interaction between two electrons, by means of phonon exchange, dominates the usual repulsive Coulomb interaction. An attractive interaction between electrons can lead to a ground state separated from excited state by an energy gap. The energy difference between the free state of the electron and the paired state appears as the energy gap at the Fermi surface. The normal electrons states are above the energy gap and the superconducting electrons states are below the energy gap at the Fermi surface. Energy gap is a function of temperature, and it is maximum at absolute zero. At $T = T_c$, pairing is dissolved and energy gap reduces to zero. Across the energy gap there are many excited states for the superconducting Cooper pairs. The critical field, the thermal properties and most of the electromagnetic properties are consequences of the energy gap.

Superconductivity occurs when such an attractive interaction between two electrons, by means of a phonon exchange, dominates the usual repulsive coulomb interaction. Two such electrons which interact attractively in the phonon field are called a Cooper pair. The energy of the pair of electrons in the bound state is less than the energy of the two states is the binding energy of the Cooper pairs. Pairing is complete at T = 0 K and is completely broken at critical temperature.

The paired electrons (Cooper pairs) are not scattered. Consequently, they can maintain their coupled motion up to a certain distance called coherence length. The penetration depth and the coherence length emerge as natural consequences of the BCS theory.

Josephson Effect

When two metals are separated by a thin (~nm)

insulating barrier, electron can tunnel from one to other. If the metals are superconducting, single electron can tunnel through the barrier unless a voltage sufficient to break up the pairs is applied. However, at zero volts, electron pairs can tunnel and give raise to a critical tunnelling current I_0. This is the Josephson effect. The tunnelling device is called a Josephson junction and forms the basis of several electronic devices.

BASIC STRUCTURAL ASPECTS OF HIGH T_c CUPRATES

The Cuprate superconductors like La-M-Cu-O, Y-Ba-Cu-O, Bi-Sr-Ca-Cu-O and Tl-Ba-Ca-Cu-O are widely studied by several groups. Holes are the majority current carries in these systems.

Superconducting Oxides

The normal state and superconducting properties are believed to arise from the strongly correlated motion of the electronic charge carriers (electrons or holes) in the CuO_2 planes that essentially define the layered cuprate HTSC. The other cations and oxygen atom in the structure provide structural stability and control the number of charge carriers in the CuO_2 planes. Literally, hundreds of novel superconducting CuO_2 planar compounds have already been synthesized, most of which are substitutional variants of about a dozen basic chemical structure. All of the high T_c oxides (i.e T_c>50 K) so far reported contain copper together with two or three other metal oxide. A brief but up-to-date account of some known superconducting materials with special emphasis on the nature of crystal structure and physical properties are given below:

La M- Cu O- System

$La_{2-x}Ba_xCuO_4$ has quasi two-dimensional structure of $K_2N_iF_4$ [13] Fig. 2.2. In this structure the transition metal ion (Cu) can interact only in the ab plane. The parent compound La_2CuO_4 and other derivatives like $La_{2-x}Sr_2CuO_4$ posses this structure with elongated Cu-Oxygen octahedral. La_2CuO_4 is orthorhombic at room temperature and becomes tetragonal around 500K. When there is an excess of oxygen it becomes superconducting around 35K. Bednorz and Muller [1] have tried to get a mixed valence state of copper in order to achieve metallic behaviour. Cu^{2+}causes an insulating compound whereas Cu^{3+} gives a non-superconducting metal $La_{2-x}M_xCuO_4$ (M=Ca, Sr and Ba) has a tetragonal structure at room temperature and becomes orthorhombic around 180K. It shows the highest T_c's of 30 and 35K for concentration around x = 0.15 and x = 0.2 for Ba and Sr. The La ion in $La_{2-x}M_xCuO_4$ can be replaced by Pr, Nd, Gd and other rare earth ions without losing superconductivity..

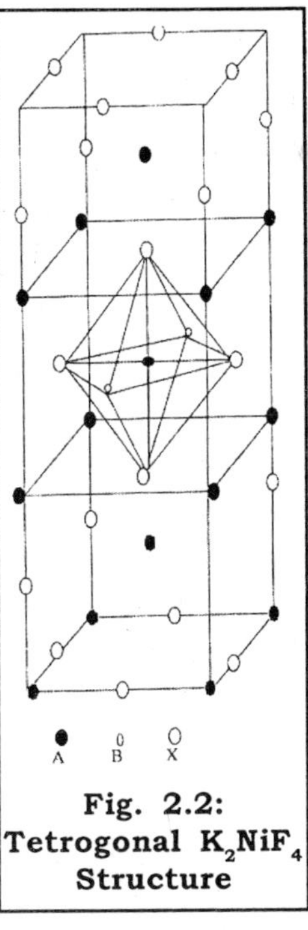

Fig. 2.2: Tetrogonal K_2NiF_4 Structure

YBCO (123) System

The $YBa_2Cu_3O_{7-\delta}$ (Fig.2.3) has an orthorhombic structure with two Cu-O sheets in the a-b plane and Cu-O chains along the b-axis [14]. The unit cell dimensions have been determined by electron and x-ray diffractions have identified the structure as being related to a cubic perovskite with one of the cubic axis a triplet. The oxygen content δ determines the lattice parameter. When δ ~ 0.6, the structure becomes tetragonal. In $YBa_2Cu_3O_{7-\delta}$ also varies with δ and the material becomes non-superconducting at δ = 0.6. Tc remains around 90K upto δ = 0.2 and then shows a plateau at 60K when δ = 0.3 – 0.4, Tc is 45K when δ = 0.5. Structural studies show that oxygen may be ordered when δ =0.5 and 0.75. In δ = 0.5 composition fully oxygenated CuO chain are present along the b-axis, alternately with fully vacant O(1) site. No clear are oxygen ordering has been found in δ =0.3 –0.4. There is a tendency for the oxygen atoms vacancies to occupy a single chain and the $YBa2Cu3O7_{-\delta}$ compounds with 0 < δ < 1 tend to have ordered arrays of completely oxygen depleted chain. When δ = 1 there is no chains.

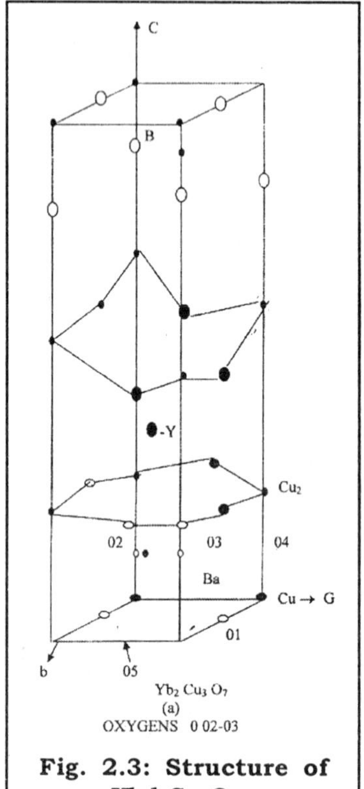

Fig. 2.3: Structure of $Ybd_2Cu_cO_{7-8}$

Bi Sr Ca Cu O Systems

From early reports that $Bi_2Sr_2CaCuO_6$ is a superconductor with a maximum T_c ~ 22K depending on the Bi/Sr ratio and subsequently bismuth cuprates of the general formula $Bi_{2-}(CaSr)_{n+1}Cu_nO_{2n+4}$ are discovered [15-17]. The T_c value ranges from 90 –100K respectively for n = 2 and n = 3. The cuprates are orthorhombic and contain Cu-O sheets, besides two

Ca, Sr-O layers and two Bi-O layers (fig.2.4). In this system T_c increases with the number of CuO2 layers. However, the usual problem encountered with then bismuth cuprates is the lack of phase purity.The divalent Ca ion in the bismuth cuprates can be replaced by trivalent rare- earth ion as in $Bi_2Ca_{1-x}Ln_xSr_2Cu_2O_{8+\delta}$ (Ln = Y or rare- earth.)

Tl Ba Ca Cu O Systems

In the thallium system, there are two series of compounds, $Tl_1Ba_2Ca_{n-1}Cu_nO_{2n+3}$ and $Tl_2Ba_2Ca_{n-1}Cu_nO_{2n+4}$ (n = 1 to 5), which can be modeled as an interleaving of rock salt and defective perovskite layers [18-19]. For example, $TlBa_2CuO_5$ can be thought of as a layer of copper oxide sandwiched between two layers of barium oxide, which is in turn sandwiched between two layers of thallium oxide. This compound, 1201, is not superconducting. A second copper layer can be inserted into the sandwich provided; it is a separated from the first by a calcium layer. This gives the 1212 compound, which has a T_c of approximately 90K. A second calcium layer allows the insertion of a third copper layer giving the 1223 compound with T_c increased to 116K. The corresponding series of compounds with T_c increased to 116K. The corresponding series of compounds with the double thallium layers are somewhat higher values of $T_{c,}$ for 2223, the highest T_c is at 128K. Thallium cuprates of the general formula $Tl_2Ca_{n-1}Ba_2Cu_{+9n}O_{2n+4}$ are tetragonal and contain two Tl-O layers and nCu-O sheets (Fig. 2.5). The T_cs of these cuprates are 80, 110, respectively for n + 1, 2 and 3. In these cuprates, there is phase in homogeneity due to co-occurrence of the different members of the series. Members of $TlCa_{n-1}Ba_2Cu_nO_{2n+3}$ families having a primitive tetragonal structure, contain a single Tl-O layer. The n + 2 and 3 members of the series shows T_cs of 90 and 115

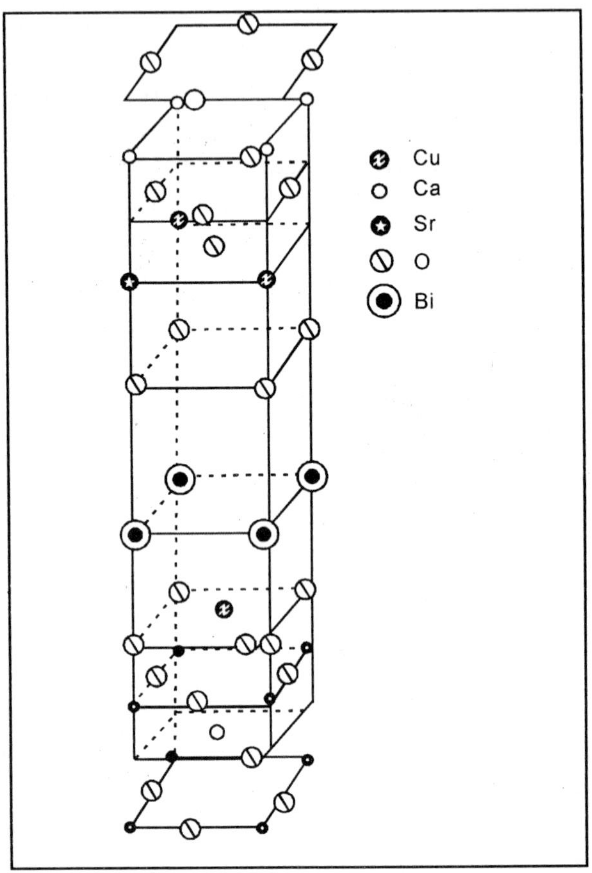

Fig. 2.4: The Crystal Structure of $B1_2CaSr_2Cu_2O_8$

respectively. Thallium cuprates with n > 3 have been characterized but T_c does not increase beyond n = 3. Partial substitution of Tl by Pb or Ca by a rare earth as in $Tl_{0.5}Pb_{0.5}CaSr_2Cu_2O_{6+\delta}$ (T ~ 90K) and $TlCa_{0.5}Sr_2Cu_2O_{6-\delta}$ (T~90K) give rise to superconductivity. Pb in this Thallium cuprateis in the 4+ state, making $Tl_{0.5}Pb_{0.5}$ $CaSr_2Cu_2O_{6-\delta}$ and $TlCa_{0.5}Ln_{+0.5}Sr_2Cu_2O_{6+\delta}$ isoelectronic. It may be noted that in the Bi and Tl cuprates as well as in the 1-2-3 compounds, there is always ca or/and Ba. At the same time superconducting Tl cuprates of

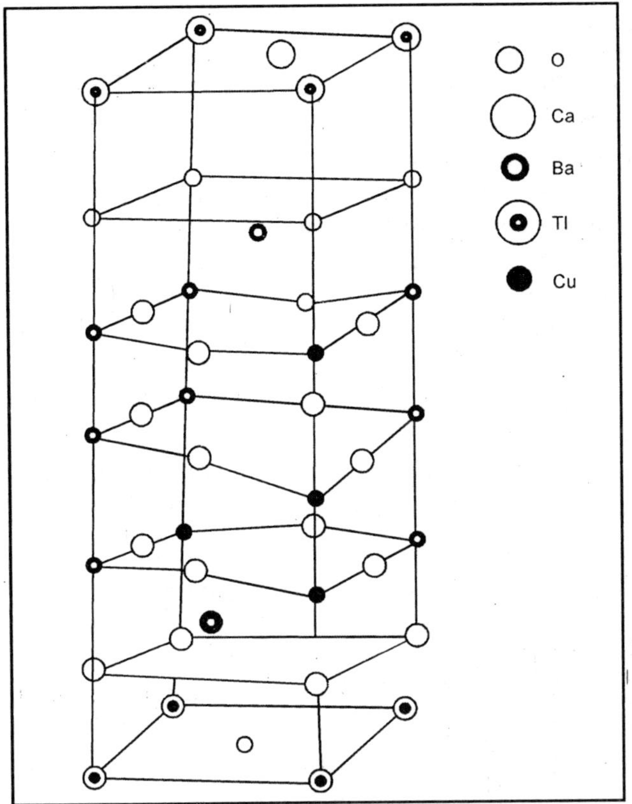

Fig. 2.5: The Crystal Structure of $TlCa_2Ba_2Cu_3O_8$

the formula $TlSr_{n+1-x}Ln_xCu_nO_{2n+3}$ have been prepared as also cuprates of the type $Tl_{1-x}Pb_xSr_{n+1}Cu_nO_{2n+3}$, which do not contain Ca or Ba or even a rare earth.

SUPERCONDUCTING STATE

Although the origin of superconductivity in the HTSC still remains unclear, the early measurements of the flux quantum as h/2e, provided that elections were paired in the superconducting state. Most experiments seems to imply zero angular momentum s-state pairing.

However, many theoretical models predict electron pairing in higher orbital momentum p or d states similar to super fluid liquid 3He and, very likely, heavy fermions superconductors also. Low temperature NMR measurements of the nuclear relaxation rate via electronic excitations exhibit a power low temperature dependence inconsistent with s-state pairing, though it is not clear whether this is an intrinsic effect. A number of research groups have looked for a rotation in the plane of polarization of light by HTSC, which could provide a signature for non-s-state pairing. How-ever, the results are confusing. It is not clear whether these differences are sample or optical - frequency dependent. The question of the symmetry of the pairing therefore remains unresolved, though a balanced s-state pairing seems most likely.

The combination of a large pairing energy $\Delta \sim KT_C$ and a small fermi velocity v_f results in an extremely small superconducting coherence length ($\zeta \sim V_f/\Delta$). For YBCO, ζ is 20-30 A in the CuO_2 planes and only (~1.5-3.0) in the perpendicular direction. Such values can be determined by the upper critical field, $Hc_2 \sim h/2e^2$ which are much higher (typically 100T than for conventional superconductivity can be destroyed by defects even on an atomic scale- by departures inexact stoichiometry, oxygen vacancies, dislocations, etc. It also results that in very small pinning energies -$\mu 40Hc^2\zeta^3$ where Hc is the thermo-dynamical critical field) for flux lines associated with flux penetration in the superconducting state. The small pinning energy means that flux lines can be relatively easily unpinned by intrinsic thermal fluctuations. This is the main challenge for high-temperature, high current power applications of HTSC. The short coherence length also creates problems because of the difficulty of retaining atomic perfection at surface, as required for the measurement of superconducting properties by the powerful characterizations tools of surface science, and across interfaces in device structures. Any degradation close to an interface may lead to the measurement of superconducting properties that are not representative of the bulk. This is the major source of irreproducibility of the short coherence length is the ubiquitous "weak link" problem. Weak-links are simply localized regions of the superconductor where the properties are degraded by structural defects-variations in chemical stoichiometry, vacancies, lattice defects, or even grain boundaries within otherwise perfect single crystals.

SPECTROSCOPIC INVESTIGATION OF HTSC

Vibration spectroscopy is the most useful and powerful technique for the elucidation of molecular structure, intra- and inter-molecular forces, structural phase transitions for qualitative analysis. As the X-ray, neutron and HREM studies provide information about the position and arrangement of atoms in the molecules whereas the vibrational spectra give an insight into the discrete motion of the atoms in the molecular system.

Electromagnetic radiation incident and interacts with it either by absorption or by scattering. If the incident radiation excite vibrational states, absorption or emission will take place. The measured shifts of Raman scattered radiation are the same as the absorption frequency observed in the infrared and have the same origin, namely the energy difference associated with fundamental molecular vibration

RAMAN SPECTROSCOPY

When a monochromatic electromagnetic radiation on a molecular system - solids, liquids and gases - the fluctuation electric field associated with the radiation interacts with molecular vibrations causing a periodic change in Polaris ability and as a result, the scattered phonon have frequency shifts characteristic of vibration energies of the molecular system. The difference in frequency between the incident and the scattered electromagnetic radiation is called Raman shift and this is independent of the exiting radiation. Hence for a molecular vibration to be Raman active, there must be a change in polarisability of the molecule.

The role played by the phonoms in the mechanism of superconductivity in the perovskite family of superconductors is not fully understood although a large series of experimental results has shown considerable coupling of some phonons to electronic excitation in these systems [20]. Optical spectroscopy and especially Raman and infrared spectroscopy revealed strong direct influences of the changes in the electronic states to the phonon at the centre of the Brillion zone. And also, HTSC in all their morphologies (ceramics, twinned and untwined single crystal, oriented and partially oriented thin films) have been investigated by Raman spectroscopy. It is well established that Raman scattering is a powerful tool to observe phonon instabilities caused by structural distortion. In $RBa_2Cu_3O_7$ systems, this technique has been employed to study oxygen vacancy effects on the Cu-O chains, phase properties by micro-analysis technique, the dependence of phonon spectra upon the oxygen content, temperature dependence of phonon peaks, energy gap, impurity phase and structural phase transition above T_c. All Raman scattering studies mentioned above had a common aim at clarifying the nature of the phonon structures of such compound, which are directly or indirectly related to the mechanism of superconductivity.

INFRARED ABSORPTION

If a vibrating molecule can induce a change in the dipole moment of the molecule, the infrared radiation will be absorbed. In infrared absorption, the oscillating electric field of the photon will exert forces tending to induce the dipole moment of the molecule to oscillate at the frequency of the photon. A transfer of energy occure when the frequency of the molecule. The intensity of the absorption band is proportional to the square of the rate of change of dipole [21].

Even though Raman scattering and infrared absorption depend upon two entirely different physical phenomenon, both the techniques provide information, since in the case of most of the high temperature superconductors, they posses centro-symmetirc structure. If the structure is centro-symmetry only, even (gerade type) modes are active in Raman scattering. From the group theoretical analysis, it was found that only those ions that have point symmetry C_{4v} or C_{2v} make a contribution to Raman Scattering. Vibrations of ions with D_{2h} or D_{4h} positional symmetry do not appear in Raman Scattering in perfect because of the alternating exclusion rule.

In general intensity, line width and frequency show considerable changes when the temperature is changed. Disorder in the crystal may be pictured by temperature dependent spectroscopy. Infrared studies mainly contribute to an understanding of one of the most important basic property namely band gap energy of this new class of superconductors. Another interesting property is the observation of anomalous phonon behaviour near T_c.

EFFECT OF TEMPERATURE ON SPECTRAL FREQUENCY

Line width and frequency of a spectral line depends on the temperature. The temperature dependence is largely due to anharmonic forces that cause the volume of the crystal to become temperature dependent (22,23). The change in the separation of atoms that results from a temperature dependent volume is analyzed by harmonic forces and this change gives rise to a shift in the frequency. This shift will move to high frequency on cooling. The presence of anharmonic term and/or linear dielectric nature that allows the phono to couple, gives rise to shift to a lower frequency cooling.

Anharmonic forces, which arise due to the motion of oxygen atom, are repulsive in nature and its presence in addition to anharmonic attractive forces causes non-linearity of the binding forces leading to a change to line width of the spectral lines. This anharmonic motion of oxygen gives rise to many exciton mechanisms. This anharmonic vibrations of oxygen either in the plane or apical position is mainly causing the pairs mechanism and for transfer of dopend hole to the CuO_2 planes. The anharmonic vibrations of this atom can give rise to a strong enough coupling to explain high T_c values [24]. The O_A seems also to be crucial to the transfer of change between the superconductive CuO_2 planes and the adjacent charge reservoir layers, i.e. CuO_2 chains in 123, TIO/Bi system. The doped holes reside mainly on the O sites. The pairing between these carries is mediated by a coupling of these carries to the anharmonic vibration of the apex oxygen. The doping is necessary both to provide the carries which finally produce the super conducting effect.

PHASE TRANSITION AND SOFT MODES

The temperature at which a solid undergoes a phase change may be called critical temperature T_c and often the phase change is connected with a transition to some ordered/disordered state. Thus the frequency of a soft mode tends to change as the temperature approaches T_c. The soft mode theory explains a continuous structural phase transition as arising from a dynamical instability of the crystal against a particular normal mode of vibration [25].

NORMAL COORDINATE ANALYSIS

Even though a fairly good amount of literature is available on the vibrational spectra of high temperature superconductors, still some specific feature in the experimental vibration spectra could not be assigned reliable to a definite type of vibration. Hence a normal coordinate analysis (NCA) which is applicable to zero wave-vector normal-mode vibrations have been carried out for the high temperature superconductors and the assignment of specific features like phonon softening and hardening were looked in for the clear understanding of the superconducting mechanism in these new class materials.

In this method the frequency of the normal vibration is determined by the kinetic and potential energies of the system. Wilson's GF matrix method [26] modifies by Shimanouchi *et al.* [27] for solids is applied for the calculation of optically active vibration frequencies. The kinetic energy is determined by the masses of their individual atoms and their geometrical arrangement in the molecules but the potential energy (PE) arises from interaction between the individual atoms described in terms of the force constants. Assuming reliable potential constants for various bonds, the vibration frequencies have been evaluated. Fine-tuning is done until the available observed frequency and the present evaluated frequency matches perfectly. Internal coordinates like bond length and bond angles are used in the kinetic energy expression. Since the potential and kinetic energies are described in terms of internal coordinates, they have clear physical meaning since these force constants are characteristics of bond- stretching and angle deformation involved.

Besides non-central forces such as involved in angle bending can be included in the normal coordinate analysis, which was not possible in the lattice dynamics. Also in NCA, Potential Energy Distribution (PED) indicates the contribution of individual force constants to the vibration energy of normal mode for the clear understanding of the specific vibration the species involved. The normal coordinate calculations were performed to support the assignment of the vibration frequencies and to obtain PED for various mode. In the normal coordinate analysis, PED play an important role for characterization of the relative contributions from each internal coordinate to the total potential energy associated with particular normal coordinate of the molecule. The contribution to the potential energy from the individual diagonal elements give rise to a conceptual link between the empirical analysis of infrared spectra of complex molecules dealing with characteristic group frequencies and the theoretical approach from the computation of the normal modes. The NCA gives complete assessment of all normal vibration modes of the system.

TOWARDS APPLICATIONS

In spite of numerous difficulties, the improvement in the superconducting properties has opened the door to a host of futuristic applications. The basic properties of the superconductors electrical have been successfully exploited for their practical applications. The three fundamental parameters that determine the economic feasibility for the applications are:

(i) critical temperature (T_c),
(ii) critical current (J_c) and
(iii) critical magnetic field (Hc_2).

The higher the above three values, the better will be the practical applications. The discovery of HTSC has unique challenges for the electric power industry. It holds the promise of a new technology, which becomes economically viable only in the area of large-scale power generation and distribution. The large-scale application of superconductors in power industry depends on two basic properties:

(i) Its ability to carry very high current with no or very little loss
(ii) Its ability to produce medium to high magnetic field

The first property is used in power transmission while the second in power generation and storage.

Power Transmission

Since superconductors exhibit zero resistance below $T_{c,}$ there is no energy dissipation (I^2R loss) associated with the flow of current through it. Hence, superconductors have been proposed for power transmission cables for loss-less transmission. Many superconductors can carry, without energy loss, 100 to 1000 times larger current densities compared to normal conductors like copper. The need for cryogenic apparatus for cooling effectively excludes them for small-scale transmission due to high costs. It becomes economically now exists for power transmission and in electronic circuitry employing liquid nitrogen as the coolant.

Power Storage

Another area, where large-scale application of superconductors are conceptualized is in the area of superconducting magnetic energy storage. Here one utilizes the properties of a superconductor to sustain very high magnetic field. With d.c. operating the current in a superconducting coil short-circuited at its ends, flows indefinitely so that superconductors coils can be used as an inductive storage device. The most attractive application of such a device will be in load balancing in existing power grids.

Superconducting Magnets

The most important application of superconductors is in magnets. The success of superconductivity started with small laboratory magnets and extended through NMR to whole body magnetic resonance medical imaging (MRI). Superconducting magnets are also used in nuclear physics for the acceleration and focusing of particle beams and in particle detecting bubble chambers.

Motor, Generators and Alternators

Homopolar dc motors and generators, and alternators have been successfully demonstrated using superconducting winding. The advantage of superconductivity in all of the above applications lies in the reduction in size for a given power over a conventional machine. The lower cost of refrigeration to nitrogen rather then helium temperatures, counted with the greater reliability of nitrogen refrigerators over helium refrigerators will inevitably make machines based on 77K superconductors more attractive economically than those using existing commercial materials.

Electronic Applications

Though current in the electronic device are small, due to restricted dimensions, current densities are similar to those in large machines. However, magnetic fields are generally small. In thin film form, ceramic superconductors can have adequate critical current densities. It seems reasonable to suggest that the first applications of ceramic superconductors will be in electronic circuits and devices. The use of superconductors, with high current densities and absence of dissipation, implies the possibility of higher packing density on a circuit board or chip. Conventional superconductivity is not very attractive in these applications, as the associated semi-conductor elements do not operate well at helium temperatures. However, nitrogen temperatures are beneficial to semiconductor operation range. Thus the possible use of ceramic superconductors as passive circuit elements can confer a double advantage. All of the above applications require the production of high current density superconducting films, which can be processed in compatibility with the semi-conducting components, a goal which has not yet been proven to be readily achievable. Larger passive electronic devices, which can profit from the use of superconductivity, include RF cavities and antennae. The absence of dissipation improves efficiency. The Meissner effect allows superconductors to be used for magnetic shielding.

The largest class of active superconducting devices is two terminal devices based on the Josephson effect.

This two-junction device, called a SQUID can be used as a sensitive detector and very accurate meter for magnetic flux. It can also be used as amplifiers, magnetometers and as digital components.

FUTURE PROSPECTS

Continuing advantages in the experimental and theoretical understanding of HTSC can be expected confidently, particularly as sample quality improves. Superconductivity at high temperatures, even at room temperature via as yet unrevealed mechanism, may still remain a possibility. The next significant breakthrough is likely to come from a totally unexpected direction, combining of serendipity, as in the latest discovery of superconductivity in the fullerenss.

REFERENCES

1. J. G. Bednorz and K. A. Muller, *Z. Phys.* **B 64**: 189 (1986).
2. C. W. Chu, P. H. Hor, R. L. Mrg and L. Gao, Z. J. Huang and Y. Q. Wang, *Phy. Rev. Lett.* **58**: 405 (1987).
3. M. K. Wu, J. R. Ashburn, C. J. Torng, P. H. Hor, R. L. Meng, L. Gao, Z. J. Huag, Y. Q. Wang and C. W. Chu, *Phy. Rev. Lett.* **58**: 908 (1987).
4. H. Maeda, Y. Tanaka, M. Fukutomi and T. Asano, *Jpn. J. Appli. Phys.* **27**: L209 (1988).
5. Z. Z. Sheng, A. M. Hermann, A. E. Ali, C. Almasan, J. Estrada, D. Dattaand R. J. Matson, *Phys. Rev. Lett.* **60**: 937 (1988).
6. M. J. Rosseinksy, A. P. Ramirez, S. H. Glarum, R. C. Haddon, A. F. Hebard, T. T. M. Palstra, A. R. Kortan, S. M. Zahurak, A. V. Makhija, *Phys. Rev. Lett.* **66**: 2830 (1991).
7. S. N. Putlin, E. Vantipov, O. Chmaissen and M. Marezio, *Nature,* **362**: 226 (1993).
8. C. W. Cu, *New Scientist* (1993).
9. Y. Tokura, H. Takagi and S. Uchida, *Nature,* **337**: 334 (1989).

10. C. Kittle, *Introduction to Solid State Physics,* John Wiley & Sons (1976).

11. S. V. Subramanyam and E. S. Rajagopal, *High Temperature Superconductors,* Wiley Eastern Ltd. (1989).

12. A. C. Rose-Innes and E. H. Phoderick, *Introduction to Superconductivity,* Pergamon Press Ltd. (1978).

13. P. Ganguly and C. N. R. Rao, *J. Solid state Chem.* **58**: 193 (1984).

14. A. P. Litvinchuk, C. Thomson and C. Cardona *Solid State Communications,* **83**: 343-347 (1992).

15. C. Michael, M. Herview, M.M. Borel, A. Gradin, F. Deslanders, J. Provost and B. Raveau, *Z Phys.* **B68**: 421 (1987).

16. A. Khurana, *Phys. Today* **41**(4): 21 (1988).

17. H. Maeda, Y. Tanaka, M. Fukutomi and T. Asano, *Jap. J. Appl. Phys* **27,** L209 (1988).

18. A. D. Kulkarni and W. de wette, J. Prade and U. Schroder W. Kress, *Physica Review* B 5451, 5457 (1990).

19. M. Krantz, H. J. Rosen, R. M. Macfarlane and V. Y. Lee *Phys. Rev.* **38**: 11962, 11965 (1988).

20. R. Li, R. Ferile,G. Jakob,Th. Hahn and H. Adrian, *Phys. Rev. Lett. 70,* 3804 (1993).

21. M. Horak and A. Vitek, *Interpretation and Processing of Vibration Spectra,* Willey Inter-Science (1978).

22. D. A. Kleinman, *Phys. Rev.* 118 (1960).

23. K. V. Krishna Rao, *Physics of solid state,* S. Balakrishna, M. krishnamurthi and R. R. Rao (Eds), Academic Press, London (1960).

24. Francois Gervais, *Physica. C.* 185-189, 2609 (1991).

25. R. Blinc and B. Zeks, *Soft Modes in Ferroelectrics and Anti-ferroelectrics,* North Holland Publishing Co, Amsterdam (1970).

26. E. B. Wilson Jr, J. C. Decius and P. C. Cross, *Molecular Vibrations,* McGraw-Hill, NY (1955).

27. T. Shimanouchi, M. Tsuboi and T. Miyazawa, *J. Chem. Phys.* 35, 1597 (1961).

CHAPTER 3

Mathematical Formula of Normal Co-ordinate Analysis of High Temperature Superconductors [La_2CuO_4 and $Pr_xH_{01-x}Ba_2Cu_{3-}O_7$]

INTRODUCTION

A molecule possesses three types of internal energy. These are electronic, vibrational, and rotational energies. Every molecule, at all temperatures, is continually executing vibrational motions. The complex, random, and seemingly periodic internal motions of a vibrating molecule are the result of the superposition of a number of relatively simple vibratory motions known as the normal vibrations or normal modes of vibration of the molecule. Each of these has its own fixed frequency. Naturally, when many of them are superposed, the resulting motion must also be periodic, but it may have a period so long as to be difficult to discern the generalized co-ordinates, each one of them executing oscillations of one single frequency, are called normal co-ordinates.

Let us write the equation of transformation to relate the original coordinates P 's with normal co-ordinates Q_1 's, by the defining equations:

$$p_i = \sum_{i=1} C_{i1} Q_1 \tag{3.1}$$

or $$p = CQ \tag{3.2}$$

where p and Q represent the single column matrices, (p) and (Q).

In order to write the equations of motion interms of normal coordinates, we shall have to calculate first kinetic energy T, potential energy V and then Lagrangian L interms of Q's. The Potential energy is

$$V = 1/2 \sum_{ij} b_{ij} p_i p_j \tag{33}$$

This is quadratic in p and therefore can be expressed as:

$$V = (1/2)\ p^T\ b_p \tag{3.4}$$

Putting for p from equation (1), we get

$$V = (1/2)\ C^T Q^T\ b\ CQ$$

$$= (1/2)\ Q^T C^T\ b\ CQ$$

$$V = (1/2)\ Q^T\ \lambda\ Q \tag{3.5}$$

Where

$$C^T\ bC = \lambda.$$

Since the expression of potential energy is again quadratic, we can write it as

$$V = (1/2) \sum_k \lambda_k Q_k^2$$

$$= (1/2) \sum_k \varpi_k^2 Q_k^2 \tag{3.6}$$

The Kinetic energy is given by

$$T = (1/2)\sum_{i,j} a_{i,j} p_i p_j$$

$$= (1/2)\ p^T\ a\ p$$

$$= (1/2)\ Q^T\ C^T\ a\ C\ Q \text{ [using Eq. (3.2)]}$$

$$= (1/2)\ Q^T\ Q$$

$$T = (1/2)\sum_{k} Q_k^2 \qquad (3.7)$$

where

$$C^T ac = 1$$

This can again be proved with the help of matrix algebra.

Therefore, Lgrangian will be

$$L = T - V$$

$$= (1/2)\sum_{k} Q_k^2 - (1/2)\sum_{k} \varpi_k^2 Q_k^2 \qquad (3.8)$$

which when used with

$$\sum_{k=1}^{S} [(\partial / dt)\{\partial L / \partial Q_k - \partial L / \partial Q_k\}] = 0$$

gives

$$\sum_{k=1}^{S} [Q_k + \omega_k^2 Q_k] = 0 \quad \text{or}$$

$$\left.\begin{array}{l} Q_1 + \omega_1{}^2 Q_1 = 0 \\ Q_2 + \omega_2{}^2 Q_2 = 0 \\ \cdots\cdots\cdots\cdots \\ \cdots\cdots\cdots\cdots \\ \cdots\cdots\cdots\cdots \\ Q_s + \omega_s{}^2 Q_s = 0 \end{array}\right\} \qquad (3.9)$$

In above equation, we find that co-ordinate Q_1 corresponds only to ω_1, Q_2 to ω_2 and so on. Thus, each coordinate executes only one single frequency oscillation and therefore $Q_{1,}$ Q_2 etc. are termed as normal coordinates.

This solution of Eq. (3.9) are

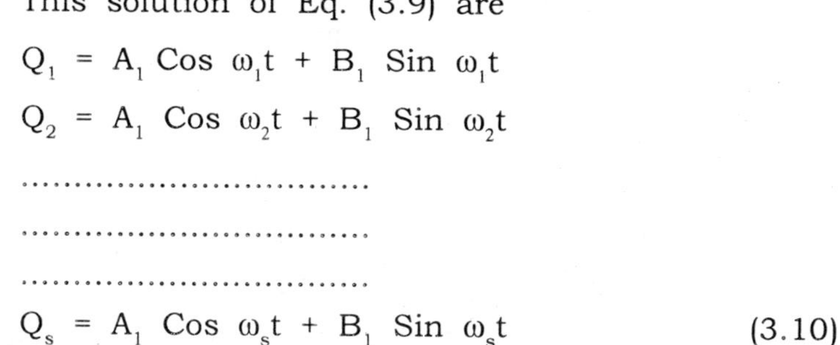

$$Q_1 = A_1 \text{ Cos } \omega_1 t + B_1 \text{ Sin } \omega_1 t$$

$$Q_2 = A_1 \text{ Cos } \omega_2 t + B_1 \text{ Sin } \omega_2 t$$

....

....

....

$$Q_s = A_1 \text{ Cos } \omega_s t + B_1 \text{ Sin } \omega_s t \qquad (3.10)$$

If we choose constants such that all except A_1 and B_1 are zero, then only normal coordinate Q_1 will vary periodically with time, while the rest will remain zero for all times. Such a situation corresponds to normal modes of vibration i.e., we say that the system is vibrating in normal mode. Obviously three will be S normal modes of vibration and S normal frequencies $\omega_{1,}$ ω_2, ω_3 . . . ω_s corresponding to each normal coordinates Q_1, Q_2, . . . Q_s.

THE SYMMETRY OF NORMAL VIBRATIONS

Consider a molecule containing N-atoms. The position of each atoms is referred by specifying three coordinates (e.g., the x, y and z Cartesian coordinates. Thus the total number of coordinate values is 3N and the molecule has 3N degrees of freedom since each coordinate value may be specified quite independently of the others. However, once all 3N coordinates have been fixed, the bond distance s and bond angles of the molecule are al. ' fixed and no further arbitrary specifications can be made. Now the molecules are free to move in three-dimensional space, as a whole, without change of shape. We can refer to such movement by noting the position of its centre of gravity at any instant. Three coordinates are needed to specify the location of the centre of gravity of the molecule, and so three of the displacements correspond to the translational motion of the molecules as a whole. The remaining 3N - 3 are non translational 'internal mode of the molecule.

In general, also, the rotation of a non-linear molecule can be resolved into components about three perpendicular axes. Specification of these axes requires three degrees of freedom, and the molecule is left with 3 N - 6 degrees of freedom. The only other motion allowed to it is internal vibration, so non-linear N atomic molecule can have 3N - 6 different internal vibrations.

Non-linear : 3N - 6 fundamental vibrations (3.11)

On the other hand, if the molecule is linear, three is no rotation about the bond axis ; hence only two degrees of rotational freedom are required, leaving 3N-5 degrees of vibrational freedom - one more than in the case of a non-linear molecule.

Linear : 3N - 5 fundamental vibrations (3.12)

In both cases, since an N-atomic molecule has N – 1 bonds between its atoms, N-1 of the vibrations are bond-stretching motions, the other 2N-5 (non-linear) or 2N-4 (linear) are bending motions. The number of normal modes is equal to the number of vibrational degrees of freedom possessed by the molecule.

For example, consider a diatomic molecule: Here N = 2, 3N - 5 = 1 and thus there can be only one fundamental vibration. Consider water molecule which is a non-linear triatomic molecule, N – 3, 3N – 6 = 3. It has thee modes of vibration (and three modes of rotation). Each motion is described as stretching or bending depending on the nature of the change in molecular shape; CO_2 is a linear triatomic molecule, and has four modes of vibration (and only two modes of rotation). Even a middle-sized molecule such as naphthalene (C_{10} H_{8}) has 48 distinct modes of vibration.

These vibrational motions are referred to as the normal modes of vibration (or normal vibrations), of the molecule; in general a normal vibration is defined as a molecular motion in which all the atoms move in phase and with the same frequency. The normal modes have two important properties:

1. Each of the vectors representing an instantaneous atomic displacement may be regarded as the resultant of a set of three basis vectors.
2. Each of the normal modes forms a basis for or "belongs to" and irreducible representation of the molecule.

There are many ways in which the set of basis vectors might be chosen, but only two are of interest. In the first method, separate Cartesian coordinate systems are attached to each of the atoms constituting the molecule. In such coordinate systems unit vector s are placed along X, Y and Z axes. The vectors representing the displacement of the its atom can be expressed as the vector sum of the displacement coordinate x_i, y_1, z_1. Thus, all 3N degrees of freedom of motion of the molecule can be represented by suitable combinations of the 3N Cartesian displacements.

The second method of resolving the displacement vectors of the normal modes is to use the basis vectors related to the internal coordinate - the inter-atomic distances and inter-bond angles - of the molecule. Normally, we choose first the changes in inter-atomic distances between bonded atoms and then as many change in bond angle as are necessary to provide a set of 3N - 6 internal displacement vectors. For example in BF_3 molecule we require 6 internal displacement vectors to represent its 6 normal modes. We may choose the changes in B-F bond distances and the changes in two of the F-B-F angels. The sixth choice could be either a change in the remaining F-B-F angle or a change in the angle between a B-F bond axis and the plane of the molecule.

CLASSIFICATION OF NORMAL VIBRATIONS

Symmetry plays an important role in the structure of the molecule. It decides which of the vibrations are permitted or executed in Raman or infrared spectra. Since molecules possess symmetry, they are classified into various groups based on symmetry elements. Each symmetry element possesses a symmetry operation.

SYMMETRY ELEMENTS

Centre of Symmetry (i)

A molecule has a centre of symmetry (i) if by reflection at that centre (inversion), the molecule is transformed into itself (or) if there exists a point within a molecule such that for every atom a straight line can be drawn through the point connecting this atom with an equivalent atom at an equal distance from the point. Then this point is called the centre of symmetry e.g., Carbon dioxide and ethylene. Molecules which do possess a centre of symmetry may or may not have an atom at the centre. It is obvious that a molecule has only one centre of symmetry.

A p-fold rotation axis of Symmetry (C p):

The molecule is said to have a p-fold rotation axis of symmetry if by rotation through 360/p yield a configuration indistinguishable from the original one. E. g., Water molecule, Ammonia molecule. Here rotation through 180° yields p = 2. Hence it has two fold rotation axis of symmetry.

Plane of Symmetry

A molecule has a plane of symmetry if by reflection at that plane, the molecule is transformed into itself. In order words a plane of symmetry bisects the molecule into two equivalent parts, on part being the mirror image of the other e.g., Water molecule. A water molecule has two vertical planes of symmetry.

Improper Rotation (S p)

A molecule has a p-fold rotation reflection axis if by rotation through 360/p followed by reflection at a plane perpendicular to the axis of rotation yields a configuration indistinguishable from the original one e.g. Methane molecule.

Identify E or I

All molecules possess the identity element. The corresponding symmetry operation is leaving the molecule unaltered. Hence, the resulting molecule cannot be distinguished from the original one.

The vibrational and rotational energies of a molecule is increased on passing IR radiation through the sample. There are two types of molecular vibrations. They are (1) Stretching and (2) Bending.

Stretching Vibration

Here the distance between the atoms increases or decreased but the atom remain in the same bond axis. There are two types of stretching vibrations. If the vibration does not alter any of the symmetry properties of the molecule, it is said to be a symmetric vibration; on the other hand, if the vibration is such that reflection of the molecule in any plane of symmetry results in a change of sign of the displacement, it is called an antisymmetric vibration. In a diatomic molecule there is only one mode of vibration, the bond stretch. In a polyatomic molecule there are several modes because bonds may stretch and angles may bend.

Bonding Vibrations

The atoms change their position with respect to the original bond axis. There are four types of bending vibrations. They are:

(1) Scissoring in which the two atoms approach each other.

(2) Rocking in which the movements of the atoms take place in the same direction,

(3) Wagging in which the two atoms move up and below the plane with respect to the central atom,

(4) Twisting is the type of vibration, one of the atoms moves up the plane while the other moves down the plane with respect to the central atom.

The bending vibration, υ_2, is also symmetric. Rotation of the stretching motion of molecule about the axis, however, produces a vibration which is in antiphase with the original and so this motion is described as the antisymmetric stretching mode. The presence of symmetry in a molecule simplifies the study of its modes of vibration. Infact, each of 3N – 6 (or 3N – 5) vibrations are symmetric or anti-symmetric with respect to any symmetry element (e.g. plane of symmetry or a centre of symmetry) of the molecule. Both symmetric and anti-symmetric modes are independent in the sense that if one is excited, then it does not excite the other. They are two of the normal modes of the molecule, its independent, collective vibrational displacements.

In general, a normal mode is an independent, synchronous motion of atoms or groups of atoms that may be excited without leading to the excitation of any other normal mode. Generally, a normal mode is a composite motion of simultaneous stretching and bending of bonds. In order to be infra-red active, there must be a dipole change during the vibration and this change may take place either along the line of the symmetry axis (parallel to it) or at right angles to the line (perpendicular). Bending modes are infrared active. This vibration is referred to as the perpendicular vibration. If there is no change in the dipole moment accompanying the vibration, then that particular vibration will be inactive in the infrared spectrum. This vibration is referred to as the parallel vibration. The symmetric stretch produces no change in the dipole moment (which remains zero) so that this vibration is not infra-red active.

The normal modes of vibration of molecules are Raman active if they are accompanied by a changing polarizability. If the normal modes of vibration leave the polarizability unchanged, they are called Raman inactive. For example, consider the H_2O molecule. The polarizability ellipsoid of water molecule may be expected to decrease in size while the bonds stretch, and to increase while they compress. While undergoing the bending motion, shape of the ellipsoid changes most. Finally, in asymmetric stretching motion υ_3, both the size and shape remain approximately constant, but the direction of the major axis changes markedly. These all three vibrations involve obvious changes in at least one aspect of the polarizability ellipsoid, and all the Raman active.

The exclusion rule also helps us to decide which modes are active:

> "If a molecule has a centre of symmetry, then Raman active vibrations are infra-red inactive and vice-versa. If there is no centre of symmetry then some (but not necessarily all) vibrations may be both Raman and Infra-red active. In CO_2 molecule, symmetric stretch is Raman active; bend and asymmetric stretch is Raman inactive."

CALCULATION OF FORCE CONSTANTS: WILSON'S F-G MATRIX METHOD

The direct method of evaluating vibrational frequencies of a molecule is to write down the potential and kinetic energies of the molecule in terms of its coordinates and obtain the normal modes of the molecule by the usual methods. The method is cumbersome even for simple molecules.

Wilson's group theoretical method better known as FG matrix method while simplifying the computational steps, given a better understanding and an insight in the structural parameters such as bond length and interbond angles. (The F matrix is related to the potential energy and G matrix to the kinetic energy of the molecule). F is a matrix of force constants and thus brings the potential energies of the vibrations into the equation. G is a matrix which involves the masses and certain spatial relationships of the atoms and thus brings the kinetic energies into the equation.

The Construction of the F- Matrix

Assuming that the atoms vibrate harmonically, the potential energy V of the molecule may be written as,

$$2V = \sum_{i,j} f_{i,j} I_i I_j \qquad (3.13)$$

where I_i and I_j are changes in internal coordinates, f_{ij} are force constants and $f_{ij} = f_{ij}$. The sum extends over all values of i and j. A term such as $f_{ii}\, Ii^2$ represents the potential energy of stretching of a bond or bending of a bond angle. It is also possible to express the potential energy in terms of symmetry coordinates as follows:

$$2V = \sum_{kl} F_{kl} S_k S_l \qquad (3.14)$$

Here $F_{k,l}$ are again force constants but pertain to the vibrations described by the symmetry coordinates S_K and S_{l}. If F_{Kl} could be expressed interms of $f_{ii,}$ the solution of the secular equation becomes easy because of its symmetry factorization. The two equations of potential energy (3.13) and (3.14) can be expressed in matrix notation. Thus,

$$2V = I' fI = S' FS \qquad (3.15)$$

I's are the elements of the column matrix I and S_k's are the elements of the column matrix S, and I' and S' are corresponding row matrices.

The relationship between the internal coordinates and the symmetry coordinates can be written in matrix form as,

$$S = UI \qquad (3.16)$$

Where U is a matrix which describes a linear,

orthogonal transformations. Equation (3.16) may be rewritten as

$$I = U^{-1} S = U2\ S' \tag{3.17}$$

where U' is the inverse matrix of U and we have

$$I = (U'\ S)2 = S'\ U \tag{3.18}$$

So equation (3.15) becomes

$$(S'\ U)\ f\ (U'\ S) = S'\ FS$$

$$S'\ (U\ f\ U')\ S = S'\ FS$$

Therefore

$$U\ f\ U' = F \tag{3.19}$$

F is a symmetric square matrix of order 3N - 6. If there are m internal coordinates and n symmetry coordinates, the U matrix will have n rows and m columns.

Construction of The G Matrix

The kinetic energy T of the vibrating molecule maybe written as

$$2T = \sum_{ij} (G^{-1})_{ij} l_i\ j_j \tag{3.20}$$

in internal coordinate, or as

$$2T = \sum_{kl} (G^{-1})_{ij} S_i S_l \tag{3.21}$$

in symmetry coordinate

The G matrix may be set up by a procedure similar to that used for the F matrix and we have

$$G = U\ gU' \tag{3.22}$$

The secular equation for the harmonic vibration of a molecule may be set up with the help of F and G matrices [15]. The frequencies of the normal vibrations are given by solving the secular equation which is of the form

$$| FG - E \lambda | = 0 \tag{3.23}$$

where E is a unit matrix and λ, which brings the frequency of vibration, υ, into the equation, is defined by

$$\lambda = 4 \lambda^2 C^2 \upsilon^2 \tag{3.24}$$

We define λ as 4 $\Pi^2 \upsilon^2$ where υ is in Sec^{-1}, or 4 $\Pi^2 C^2 \upsilon^2$, where υ is in Cm^{-1}. If the G elements are expressed in atomic mass units (amu), then

$$\lambda = \frac{4 \pi^2 C^2 \nu^2}{N_0} \tag{3.25}$$

Where υ is in Cm^{-1}

$$\upsilon (Cm^{-1}) = 1303.16 (\lambda)^{1/2} \tag{3.26}$$

Equation (3.24) may be written

$$\lambda = 5.8890 \times 10^{-2} \upsilon^2 \tag{3.27}$$

The force constants f, their combination forming the elements of the F matrix is critical in their values. An appropriate combination of these values picturises the force field in the molecule. Once the elements of the F and G matrices are known for the molecule in question, the determinantal Eq. (3.23) may be written out explicitly. This is the secular equation for the vibrational problem. The great virtue of the F and G matrix method is that it affords a convenient and systematic means of employing the symmetry properties of the molecules to achieve maximum factorization of the secular equation.

The Potential - Energy Distribution

When the solutions of the secular determinant have been obtained, it is necessary to assign the calculated frequencies to the various modes of motion.

Morino and Kuchitsu have shown that the potential energy distribution rather than the normalized amplitudes is a more satisfactory quantity to use in band assignments; for a vibration of frequency υ_a, associated with a normal coordinate Q_a the potential energy of the molecule is given by

$$2V = Q_a^2 \sum_{ij} F_{ij} L_{ia} L_{ja} \qquad (3.28)$$

Such terms and large only when i = j, since the diagonal force constants F_{ii} are much greater than the off-diagonal constants F_{ij} consequently we need calculate only the terms of the form

$$F_{ii} L_{ia}^2$$

Since the normalization condition

L'FL = λ gives relations of the form

$$\sum_i F_{ij} L_{ia} L_{ja} = \lambda_a \qquad (3.29)$$

Neglect of cross terms gives

$$\sum_i F_{ii} L_{ia}^2 = \lambda_a \qquad (3.30)$$

Therefore, we get a normalized potential - energy distribution, whose terms are

$$V_{ia} = \frac{F_{ii} L_{ia}^2}{\lambda_a} \qquad (3.31)$$

where V_{ia} is the contribution of the ith symmetry coordinate to the potential energy of the vibration whose frequency is $\upsilon_{a.}$

If one term in the potential energy distribution, $V_{i\,a}$, is much greater than any of the other terms $V_{j\,a}$, V_{ka}, . . ., we are justified in regarding the mode V_a as a pure vibration of the ith coordinate; if two or more terms are roughly equal, we must regard the vibration as coupled [1-18].

REFERENCES

1. S.D. Rose, *Inorganic Infrared and Ramnd Spectra*, McGraw-Hill Book Company, England, 1972.
2. S.L. Gupta, V. Kumar, R.C. Sharma, *Elements of Spectroscopy*, Pragati Praksahan, India, 1983.
3. R.T. Conley, *Infrared Spectroscopy*, Allyn & Bacon Ltd., 1966.
4. G. Herzberg, *Molecular Spectra and Molecular Structure:* Vol. 2, *Infra-red and Raman Spectra of poly atomic molecules*, Van Nostrand, USA, 1945.
5. K. Nakamotao, *Infrared Spectra of Inorganic and Co-ordination Compounds*, John Wiley, New York, 1st edition, 1963.
6. C.N. Banwell, *Fundamentals of Molecular Spectroscopy*, Tata McGraw-Hill Book Co., New Delhi, Second Edition 1975.
7. S.E. Wiberley, N.B Colthup and L.H. Daly, *Introduction to Infra-red and Raman Spectroscopy*, Academic Press, New York, 1st edition, 1964.
8. W.A. Bingal, *Theory of Molecular Spectra*, Wiley, 1969.
9. F. Albert Cotton, *Chemical Applications of Group Theory*, Wiley Eastern Ltd., Second Edition 1971.
10. T. Y. Wu, *Vibration Spectra and Structure of Polyatomic Molecules*, National University of Peking, Kun-ming, China, 1939.
11. J. C. Decius, *J. Chem. Phy.* 22: 1941 (1954).

12. E. Bright Wilson Jr., J.C. Decius and P. C. Cross, *Molecular vibration*, McGraw-Hill Book Company, New York, First Edition (1955).

13. E.B. Wilson, *J. Chem. Phys.* 7: 1047 (1939).

14. P.G. Puranik, *Group Theory Applications to Molecular Vibrations*, S. Chand & Company Ltd., First Edition, 1979.

15. Y. Marino and K. Kuchitsu, *J Che. Phy.* 20: 1809 (1952).

16. E. Bright Wilson, *J. Chem. Phy.* 9: 76 (1941).

17. D. Schonland, D Van Nostrand, *Molecular Symmetry*, London, McGraw-Hill Book Company, 1st Edition (1965).

18. S. Mohan and A. Sudha, *Pramana J. Phys.* 37, 4, 327, (1991).

CHAPTER 4

Normal Co-ordinate Analysis of La_2CuO_4 and $PrH_{01-x}Ba_2Cu_3O_7$

INTRODUCTION

The HTSC marks the beginning of a new era of material science. Researchers are looking beyond the simple metals. Their alloys, binary and ternary compounds to the almost limitless range of complex molecular solids, many of which will be based on the rich chemistry of the transition metals. These materials can be expected to have interesting and technologically important electronic and magnetic properties as the familiar materials of modern day technology. The important legacy has been the major investment in new research equipment and the powerful range of theoretical, experimental and material processing techniques developed primarily for HTSC but equally applicable to a much wider range of materials of potential importance to future material scientists.

The introduction of superconductivity above 30 K in CuO-pervoskites by Bednorz Muller[1] initiated tremendous efforts in solid state physics and material sciences with the aim to isolate the phases which are responsible for the superconductivity and to search for other substances beside the La_2CuO_4 family which exhibit this phenomenon. These activities succeeded in the discovery of superconductivity in $YB_2Cu_3O_7$ at 92 K by Wu *et al.*[2] and the observation of a transition temperature above 100 K by Sheng *et al.*[3].

Parallel to the preparation of new materials a huge number of investigations have been performed in order to illuminate the nature of the superconductivity. One direction of experiments has the intention to find out what is the contribution of lattice vibrations to the superconductivity. Neutron scattering, as the method to determine the vibration spectrum throughout the whole Brillouin zone, could at the beginning of these activities only measure the phonon density of states, as shown by Ramirez *et al.*[4] Renker *et al.*[5,6], Bruesch *et al.*[7], Burer *et al.*[8], and Belushkin *et al.*[9], because large single crystals were not available. Raman and infrared (IR)-spectroscopy, however, can yield at least some of the phonon frequencies, namely the long wavelength optical phonons at the centre of the Brillouin zone, from polycrystalline materials. Even tiny single crystals embedded in polycrystalline samples can be investigated in Raman experiments with the use of microscope.

The published results of experimental work on Raman and IR-spectrosopy have been published in many papers. They cover detailed information about the lattice vibrations of the superconducting materials and their dependence on oxygen contents, element substitution and impurity phases, which appeared in various samples and partly yielded controversial results. In addition, the temperature variations of the optical phonon spectrum and the superconducting gap have been investigated.

It is the intention of this review to summarize the literature about the Raman and IR-experiments. The author hopes to have considered every article in these fields which have appeared in the first two years of investigations on the La_2CuO_4 and $YB_2Cu_3O_7$ series. Regarding the enormous number of papers, however, one or the other may have been overlooked, and the author apologies to all colleagues whose works have been omitted here unintentionally.

This work is organized as follows: First, after the discussion of the structure and the symmetry of the vibrational modes in the La_2CuO_4 compounds the Raman and IR-results are shortly summarized. Secondly, the articles on the $YB_2Cu_3O_7$ family are reviewed starting again with structural and symmetry considerations. The two sections, Raman and IR-spectroscopy on these compounds, have both been subdivided in sections dealing with the impurity phases, the influence of oxygen stiochiometry, substitution of elements, single crystals and films, the temperature dependence of the vibrations and the superconducting gap, and, shortly, with the results from magnetic excitations. The following third sections summarize the lattice a dynamics calculation performed for both compounds and compares the results with the experimental findings. Finally the isotope effect is discussed from which information about the relevance of the lattice vibrations for the superconductivity had been expected.

La_2CuO_4 and the related Sr and Ba-doped compounds have been after their discovery been subject of spectroscopic investigations [10-20]. Yet the number of published articles on these materials is rather small compared to the investigations reported for $YB_2Cu_3O_7$. This may have been due to the strong attraction which this new 92 K material has gained and/or experimental difficulties with the La-compound as mentioned below.

NORMAL COORDINATE ANALYSIS OF ZERO WAVE VECTOR VIBRATION OF La_2CuO_4

The stoichiometric compound La_2CuO_4 crystallize at high temperatures in the K_2NiF_4 structure (D^{14}_{4h}-I4/mmm) with one chemical formula unit per unit cell (Jorgensen *et al.*[21]). At lower temperatures the crystals undergo a structural phase transition into an orthorhombic phase (D^{18}_{2h}-Cmca) by doubling the unit cell in a √2a x √2a fashion. The exact transition temperature depends on the oxygen stoichiometry of the sample and occurs between 430 K and 530 K (Johnsoton *et al.*[22]). Sometimes the structure is also denoted as (D^{18}_{2h}-Bmab). Both assignments may be transferred into each other changing the labels from the orthorhombic axes from *abc* to *acb*, respectively [23]. Table I give the results of an analysis of the phonon symmetries at the Γ-point of the reciprocal lattice. In the high temperature phase four Raman-active and seven infrared active modes should be observable. Below the structural phase transition the number of Raman and IR-modes increases because the number of atoms in the unit cell is doubled. 18 Raman and 21 IR active modes are expected. Some of them can be deduced from the zone centre modes of the tetragonal phase, the others from zone boundary modes now becoming zone centre modes because of the cell doubling.

Doping La_2CuO_4 with Sr (or Ba) shifts the tetragonal orthorhombic transition to lower temperatures and yields the superconducting material $(La_{1-x}Sr_x)_2CuO_4$ with the highest transition temperature in this series of 40 K for x ~ 0.075.

The interest in the high-T_c superconducting cuprate-crystal is stimulated, on the one hand, by the possibility of phonons taking part in carrier pairing processes and, on the other hand, by peculiar dynamic properties of these complex compounds with a layered structure. It is not surprising; therefore, that many papers report studies of thermal excitations of the crystal lattice of high T_c superconducting cuprates and related compounds. In particular, the La_2CuO_4 crystal lattice oscillations are experimentally studied by means of neutron [24-27] and optical [28-30] spectroscopy. The results of inelastic neutron-scattering experiments giving information about phonon dispersion relation [27] or phonon spectral density [24-26] are usually interpreted in terms of collective oscillating excitations in each of which all the atoms take part. The information about the oscillations of individual atoms of the compound is thus lost.

Analysis of the literature shows that two approaches are used to get information about oscillating properties of single atoms of a complex compound. The first is based on model calculations of crystal-lattice dynamics, and the second on the analysis of indirect information about thermal atomic motions, obtained by neutron recoil spectroscopy for neutrons of energy ≈1 eV[10] by resonance neutron absorption and by neutron diffraction. All these methods suffer from an important short coming, namely the characteristics of thermal atomic motion are determined as fitting procedure parameters, together with a lot of other parameters, which, in some cases, can substantially affect the final result. Moreover, neutron diffraction data do not allow distinguishing between static and dynamic atomic displacements from equilibrium in the crystal lattice. Apparently, these very difficulties account for a large spread in data concerning the mean oscillating energies of single atoms in La_2CuO_4 and their effective Debye temperatures. For example θ_D for copper atoms is 1500 K, according to (Ikeda *et al.* [31]) while in (Mook *et al.*[32]) it is 430 K.

The most reliable information about thermal motion of a definite type (i) of atoms in a complex compound can be obtained from the partial oscillating spectrum

$$g_i(E)=g(E)\langle |e_i|^2 \rangle,$$

where e is the polarization vector of the *i*th atom when the latter oscillates with energy E. Thus, the partial spectrum is determined by superposition of all normal modes with weight allowing for the *i*th atom displacement in lattice oscillations with energy E.

As shown in Parshin *et al.* [33] and Soldatov *et al.* [34], the method of isotopic contrast in inelastic neutron scattering makes it possible to experimentally restore the partial oscillating spectra of single atoms. In the present study this method was used to investigate the oscillations of Cu, La, and O atoms in La_2CuO_4.

The study of normal coordinate analysis and the free carriers is important for the understanding of the physical nature of high temperature superconductors. Raman and far-infrared studies of these superconductors have contributed significantly to the understanding of new class of superconductors. Cardona and coworkers [35] studied the infrared and Raman spectra of the superconducting cuparate perovskites $MBaCu_2O_2$ (M = Nd, Er, Dy, Tm and Eu) and reported the possible origins of phonon softening and the systematic variation of phonon frequencies with the ionic radius. Here an attempt has been made to perform the normal coordinate analysis for the phonon frequencies and the form of the zero wave vector vibrations for the La_2CuO_4 superconductors.

The high Tc superconductor La_2CuO_4 System crystallizes in the body-centered tetragonal (bct) system, which belongs to the space group 14/mmm (D^{17}_{4h}). The simple tetragonal (bct) unit cell of La_2CuO_4 and the numbering of the atoms are shown in Fig. 4.1. The 6 atoms of the unit cell yield a total of 14 optical vibrational modes. All the above calculations are made at q = 0. Once of A_{2u} and E_u modes corresponds to acoustic vibrations with frequency ù = O. These normal modes are distributed as follows:

$A_{1g} + A_{2u} + E_g + E_u$	from the motion of 2 La atoms
$A_{2u} + E_u$	from the motion of Cu atoms
$A_{2u} + B_{2u} + 2\ E_u$	from the motion of O(1) atoms
$A_{1g} + A_{2u} + E_g + E_u$	from the motion of O(2) atoms

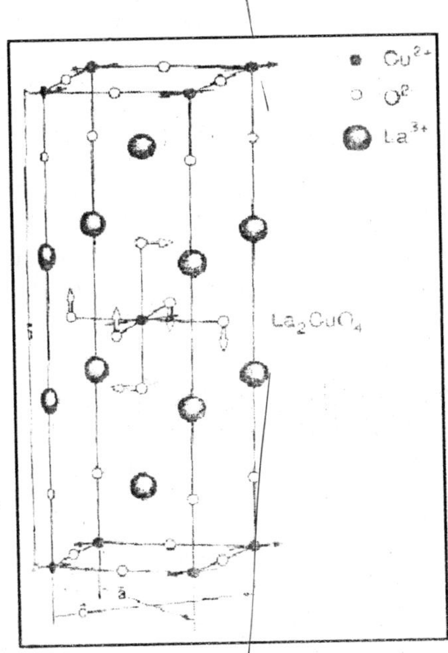

Fig 4.1 Structure of La_2CuO_4

Subtracting the translation modes $A_{2u} + B_{2u} + E_u$ the q = 0 optical modes involved in an irreducible representation are as follows:

$\Gamma_{opt} = 2A_{1g} + 2E_g + 4A_{2u} + 5E_u + B_{2u}$

The species belonging to A_{1g} and E_g Raman active modes whereas A_{2u} and B_u are infrared active modes. The A_{2u} and A_{1g} modes involve displacement along crystallographic c-axis, the B_{2u} and E_g modes along the b-axis and E_u modes along the a-axis. The normal coordinate calculation was performed using the programs GMAT and FPERT given by Fuhrer *et al.* [36]. The general agreement between the evaluated and observed normal frequencies of La_2CuO_4 is good. The calculated force constants using the above programs are given in Table 4.1. It is interesting to note that the evaluated frequencies given in Table 4.2, agree favourably with the experiment values.

To check whether the chosen set of vibrational frequencies makes the maximum contribution to the potential energy associated with the normal coordinate frequencies of the super conducting material, the potential energy distributions was calculated using the equation.

$$PED = (F_{ij} \ L^2_{ik}) / \ \lambda_k$$

where PED is the combination of the i-th symmetry coordinate to the potential energy of the vibration whose frequency is $V_k F_{ij}$ are potential constants, L_{ik} are L matrix elements and

$$\lambda_k = 4\Pi^2 C^2 \upsilon_k^{\ 2}.$$

RESULT AND DISCUSSION

Normal Coordinate Analysis of La_2CuO_4

The G-matrix elements have been calculated from the equilibrium geometry. The initial force constants were taken from the related molecules. The final sets of potential constants provide the stability of the crystal in relation to all vibrational modes. The vibrational frequencies and potential energy distribution values are presented in this work. The potential energy distribution indicates the contribution of an individual force constant to the vibrational energy of normal modes. It clearly indicates that there is mixing of the internal displacement coordinates.

The evaluated frequencies using the normal coordinate analysis method listed in Table 4.2 agrees favourably with the calculated normal coordinate analysis frequencies and observed experimental frequencies.

The calculated Raman phonon frequency at 218 cm^{-1} in A_{1g} symmetry is due to the vibration of La atom. The highest calculated phonon frequency at 553 cm^{-1} is due to the vibration of O(1) atoms and it is due to the stretched vibration of Cu-O(2) atoms

The lowest calculated phonon frequency at 105 cm^{-1} in Eg symmetry mode is due to the vibration of La atom and is due to the stretched vibration of O(1) –O(2) atoms

The calculated Raman phonon frequency at 333 cm^{-1} in Eg symmetry is due to the vibration of Cu atom and it is due to the stretched vibration between La and Cu atoms. The lowest calculated infrared phonon frequency at 163 cm^{-1} in A_{2u} symmetry is due to the vibration of La atoms and it agrees very well with the experimental values at 169 cm^{-1} and it is due to the stretched vibration of La–Cu atom. The calculated infrared phonon frequency at 305 cm^{-1} is due to the vibration of Cu atom and it is due to the bending vibration of O(1)-Cu-O(2) atom: The highest phonon frequency at 549 cm^{-1} in A_{2u} symmetry is due to the vibration of O(2) atoms and it agrees very well with the experimental values at 569cm^{-1} and it is due to the bending vibration of O(2)-Cu-O(3) atom.

Table 4.1: Force Constants for $LaCuO_4$ [in units of 10^2 Nm^{-1}(Stretching) and $10^{-18}Nm\ rad^{-2}$ (bending)]

Force	Bond Type	Distance (O)	Initial Value
f_o	Cu-o(1)	1.945	1.41
f_o	Cu-O(2)	1.827	1.40
f_o	La-O(1)	1.625	1.49
f_d	La-O(2)	1.723	0.80
f_e	La-O(3)	1.732	1.00
f_g	La-La	1.852	0.80
f_n	Cu-La	4.30	0.40
f_k	O(1)-O(2)	1.99	0.08
f_l	O(2)-O(3)	1.98	0.57
f_m	O(2)-O(3)	3.41	0.5
f_α	O(1)-Cu(1)-O(2)	-	1.32
f_β	O(1)-Cu(1)-O(3)	-	1.30

The lowest calculated phonon frequency at 120cm^{-1} is in E_u symmetry is due to the vibration of La atom and it agrees well with the experimental frequency at 135 cm^{-1} and it is due to the stretched vibration of La-O(1) atoms.

The next calculated infrared phonon frequency at 179cm^{-1} in E_u symmetry is due to the vibration of Cu atom and it agrees well with the experimental frequency at 196 cm^{-1} and it is due to the vibration of Cu–O(1) atom. The next calculated phonon frequency at 263cm^{-1} is due to the vibration of O(1) atom and it agrees very well with the experimental values at 283 cm^{-1}. Here the La atom vibrates at 180^0 degree out of phase to Cu-O(1) atoms. The highest calculated phonon frequency at 547cm^{-1} in this symmetry is due to the vibration of O(2) atoms and it is due to the stretched vibration of O(2)-O(3) atoms .The calculated phonon frequency at 206cm^{-1} in B_{2u} symmetry is due to the vibration of O(1) atom.

Table 4.2: Calculated Phonon Frequencies of La_2CuO_4 (Values in the parentheses are experiment frequencies)

Symmetry Species	Using Normal Co-ordinate Analysis(Cm^{-1})	Potential Energy Distribution (%)
A_{1g}(Raman)	218	$f_c(57)$, $f_e(20)$, $f_g(18)$
	553	$f_b(70)$, $f_a(11)$, $f_\beta(10)$
E_g	105	$f_k(74)$, $f_e(30)$
	333	$f_n(70)$, $f_m(21)$, $f_\beta(14)$
A_{2u}	163(169)	$f_n(81)$, $f_a(12)$
	305(498)	$f_\alpha(74)$, $f_a(19)$, $f_\beta(15)$
	549 (569)	$f_b(59)$, $f_a(30)$,$f_m(16)$
E_u	120(135)	$f_c(70)$, $f_k(20)$, $f_\beta(14)$
	179(196)	$f_a(65)$,$f_e(22)$
	263(283)	$f_a(49)$, $f_d(17)$, $f_e(21)$
	547(617)	$f_\alpha(60)$, $f_m(22)$, $f_k(11)$
B_{2u}	206	$f_g(40)$, $f_d(24)$, $f_b(12)$

NORMAL COORDINATE ANALYSIS OF THE ZERO WAVE VECTOR VIBRATIONS OF $Pr_xH_{01-x}Ba_2Cu_{3-}O_7$

Since the discovery of high T_c superconductor $YB_2Cu_3O_{7-6}$, elemental substitution has been used to make clear the machanism of high Tc superconductivity. Early studies showed that the replacement of Y site in $YB_2Cu_3O_7$ by the lanthanide elements except Ce, Tb, and Pr did not destroy its superconductivity. Only $PrBa_2Cu_3O_7$ compound which is isostructural to $YBa_2Cu_3O_7$, is neither metallic nor superconducting. Various explanations have been given for such suppression of the superconductivity [37-39]. But this suppression by Pr has not been understood yet.

Systematic studies of the lattice vibrations and the free carriers are important for understanding the physical nature of the high Tc superconductors. Recently, anomalous behaviour of the phonon band vibrating in the c-axis direction has attracted much interest [40-43], and strong coupling theory of phonon and free carriers has been reported [44]. The study of the phonon of the $Pr_xLn_{1-x}Ba_2Cu_3O_{7-y}$ systems (Ln is a lanthanide element) are also important. These systems have similar orthorhombic crystal structures with the oxygen composition almost 7, however with the increase of the Pr concentration Tc is decreased and finally superconductivity disappear. Free carrier concentration for $Pr_xLn_{1-x}Ba_2Cu_3O_{7-y}$ systems has been estimated by the fitting of the reflection spectra calculated using Lorentzian oscillators and a Drude term to the experimentally-obtained reflection spectra [45].

In this report Raman scattering and infrared (IR) reflection of $Pr_xHo_{1-x}Ba_2Cu_3O_7$ system is studied. The lattice dynamical calculations on high T_c superconductors $Pr_xHo_{1-x}Ba_2Cu_3O_7$ system are also studied using the shell model in order to understand the characteristic shifts of phonon frequencies observed by Raman and IR measurements with the substitution of Pr for Ho.

Systematic studies of the lattice vibrations and the free carriers are important for understanding the physical nature of high T_c superconductors. Recently, anomalous behaviour of the phonon band vibrating in the c-axis direction has attracted much interest, and strong coupling theory of phonon and the free carriers has been reported. In this paper infrared reflection spectra of $Pr_xLn_{1-x}Ba_2Cu_30_{7-y}$ (Lan=Lanthanide) are studied. Optical constants are calculated by the Kramers-Kronig analysis and anomalously large oscillator strength is denied for the "Barium mode" at

154 cm^{-1} at low temperatures. The origin of this anomalously large oscillator strength at low temperature is attributed to the increase of the broad background reflection due to the increase of superconducting gap in the low temperature range. Free carrier concentrations for these systems are estimated by the fitting of the reflection spectra with the calculated one from the model of the oscillators and the Drude term.

The discovery of superconductivity with transition temperatures (T_c) in the 90-K range in $YB_2Cu_3O_{7-6}$, has generated a great deal of interest in the oxide superconductors. The substitution of Y by trivalent rare-earth elements, with the exception of Ce, Pr, Pm, and Tb, yields a superconducting phase with a T_c almost identical to the $YB_2Cu_3O_{7-8}$, compound. Samples of $Rba_2Cu_3O_{7-8}$ with R = Ce and Tb prepared by the standard solid-state reaction technique yield multiphase materials consisting of $BaCeO_3$ or $BaTbO_3$, CuO, and $BaCeO_2$ which are not superconducting. No investigations have been reported for the $PmBa_2Cu_3O_{7-8}$ compound because the Pm nucleus is radioactively unstable. The $Y_{1-x}Pr_xBa_2Cu_3O_{7-8}$ system is particularly interesting since it is isostructural to the $YBa_2Cu_3O_{7-8}$ superconductor, yet the superconductivity is strongly suppressed as a function of Pr concentration. This quenching of the superconducting state is not understood in detail. Extensive measurements havc been carried out as a function of Pr concentration which includes magnetization, heat-capacity, thermo power, Hall Effect, neutron-diffraction, pressure effects, x-ray-absorption, and Raman spectroscopy. The effect of the Pr ion on the superconducting properties may help our understanding of the interplay between magnetism and superconductivity and provide insight as to the origin of the superconductivity in the high-T_c oxides.

One problem in studying the $Y_{1-x} Pr_xBa_2Cu_3O_{7-8}$ system is its tendency to phase separate into $YBa_2Cu_3O_{7-8}$ with T_c = 90 and $Y_{1-x}Pr_xBa_2Cu_3O_{7-8}$ of varying *x* value and a reduced T_c. This type of phase separation can be clearly seen in the filed-cooled Meissner measurements, while not showing up at all in the powder x-ray diffraction patterns. The resistively data on these phase separated samples showed broad transitions. It has been overcome this problem through specific annealing conditions of time, temperature, and atmosphere.

It has been reported in the earlier literature that the preparation, structure, oxygen, content, resistivity, magnetization, and critical field in the single phase $Y_{1-x}Pr_xBa_2Cu_3O_{7-8}$ system. The electronic coefficient of specific heat, ϒ, the density of states, *N(O)*, and the Ginzburg-Landau parameters ξ_{GL}, λ_{GL},and κ_{GL} are estimated along with the exchange interaction parameter, from the measured temperature dependence of H_{c2} and the Pauli susceptibility.

The high T_c superconductor $Pr_xH_{01-x}Ba_2Cu_{3-}O_7$ System crystallizes in the simple tetragonal (st) system, which belongs to the space group $P4/_{mmm}(D'_{4h})$ The simple tetragonal (st) unit cell of Pr_x H_{01-x} Ba_2 $Cu_{3-}O_7$ and the numbering of the atoms are shown in Fig 4.2. The 14 atoms of the unit cell yield a total of 36 optical vibrational modes. All the above calculations are made at q = 0. Once of B_{1u} and E_{3u} modes corresponds to acoustic vibrations with frequency ω = 0. These normal modes are distributed as follows.

$B_{1u}+ B_{2u}+ B_{3u}$ from the motion of Pr atom

$B_{1u}+ B_{2u}+ B_{3u}$ from the motion of Ho atom

$A_{1g} +B_{2g}+B_{3g}+B_{1u}+B_{2u}+ B_{3u}$ from the motion of 2Ba atom

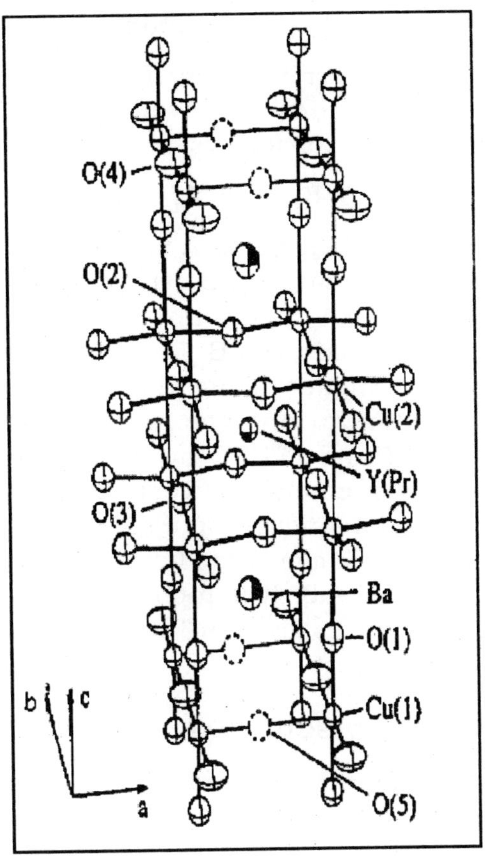

Fig 4.2: Structure of $Pr_xH_{01-x}Ba_2Cu_3O_7$

$A_{1g} + B_{2g} + B_{2g} + B_{1u} + B_{2u} + B_{3u}$	from the motion of 2Cu atom
$B_{1u} + B_{2u} + B_{3u}$	from the motion of 9Cu atom
$2A_{1g} + 2B_{2g} + 2B_{2g} + B_{1u} + B_{2u}$	from the motion of 4(1) atom
$A_{1g} + B_{2g} + B_{2g} + B_{2u} + B_{3u}$	from the motion of O(2) atom

$B_{1u} + B_{3u}$ from the motion of O(3) atom

Subtracting the translation modes $B_{1u}+B_{2u}+B_{3u}$ the q = 0 optical modes involved in an irreducible representation are as follows.

$$\Gamma_{opt} = 5\ B_{2g} + 5\ B_{3g} + 5\ A_{1g} + 8\ B_{1u} + 7\ B_{2u} + 8\ B_{3u}$$

The species belonging to A_g+B_{3g} and B_{2g} are Raman active modes whereas $B_{1u}+B_{2u}+B_{3u}$ are infrared active modes. The B_{1u} and A_g modes involve displacement along crystallographic c-axis, the B_{2u} and B_{2g} modes along the b-axis and B_{3u} modes along the a-axis. The normal coordinate calculation was performed using the programs GMAT and FPERT given by Fuhrer *et al* [36]. The general agreement between the evaluated and observed normal frequencies of $Pr_xH_{01-x}Ba_2Cu_{3-}O_7$is good. The calculated force constants using the above programs are given in Table 4.3. It is interesting to note that the evaluated frequencies given in Table 4.4 agree favourably with the experiment values.

RESULTS AND DISCUSSION

Normal Co-ordinate Analysis of $Pr_xH_{01-x}Ba_2Cu_{3-}O_7$

The evaluated frequencies using the normal coordinate analysis method listed in Table 4.4 agrees favourably with the calculated lattice dynamical frequencies and observed experimental frequencies.

The lowest calculated Raman phonon frequency at 290 cm^{-1} is due to the vibration of Pr atoms it agrees very well with the observed values at 360 cm^{-1}. The calculated Raman phonon frequency at 330cm^{-1} is due to the vibration of the Ho atoms and it agrees with the observed phonon frequency at 340 cm^{-1} and it is due to the stretched vibration of Ho-O(1) atoms. The calculated Raman phonon frequency at 355cm^{-1} is due to the vibration of Ba atom and agrees with the observed phonon frequency at 360 cm^{-1}. And it is due to the stretched vibration of Ba-Cu atoms. The calculated Raman phonon frequency at 400cm^{-1} is due to the vibration Cu atom and it is due to the bending vibration of Ca(1)–Ba-Ca(2) atom. The highest calculated Raman phonon frequency at 502cm^{-1} in this symmetry is due to the vibration of O(1) atom and it is due to the stretched vibration of Pr-O(1) atom. The observed value at 500 cm^{-1} agrees very well with the calculated Raman phonon frequency.

The calculated Raman phonon frequency at 300 cm^{-1} in B_{2g} symmetry is due to the vibration of Pr atom and it agree very well with the observed value at 320 cm^{-1} and it is due to the bending vibration of O(1)-Pr-O(2) atoms. The calculated Raman phonon frequency at 480cm^{-1} is due to the vibration of Pr-O(1) atoms and Cu atom in 180^{0} out of phase to Ho-O((1) atoms and it agrees very well with the experimental value of 420 cm^{-1}. The calculated Raman phonon frequency at 450cm^{-1} is due to the vibration of Ba atom and it is due to the bending vibration of O(1)-Ba – O(2) and it agrees very well with the experimental value at 480cm^{-1}. The highest calculated Raman phonon frequency in this symmetry is 530cm^{-1} and it is due to the vibration of O(1) and it is due to the stretching vibration of Cu-O(1) atom.

The calculated Raman phonon frequency at 290 cm^{-1} in B_{3g} symmetry is due to the vibration of Pr atom and agrees very well with the experimental values at 300cm^{-1}.And it is due to the stretched vibration of Pr-O(1) atoms. The calculated Raman phonon frequency at 415cm-1 and it is due to the bending vibration of Pr-O(1) atoms. The Raman phonon frequency in this symmetry in 535cm^{-1} and it is due to the vibration of

O(1) atom and it is due to the bending vibration of O(1)-Pr-O(1) atoms.

Table 4.3: Force Constants for $Pr_x H_{01-x} Ba_2 Cu_3O_7$ (in units of 10^2 Nm^{-1}(Stretching) and 10^{-18}Nm rad^{-2} (bending).

Force Constants	Bond Type	Distance (A°)	Initial Value Constants
f_a	Pr-O(1)	2.452	2.41
f_b	Pr-Cu	3.265	3.258
f_c	Cu-O(1)	1.976	1.982
f_d	Cu-O(2)	2.43	2.43
f_e	Cu-O(3)	2.45	2.45
f_g	Ba-O(1)	2.956	2.98
f_h	Ba-O(2)	2.811	2.810
f_i	Ba-O(3)	2.594	2.596
f_j	Ba-Cu	3.389	3.381
f_k	Ho-O(1)	2.12	2.11
f_l	Ho-O(2)	3.18	2.08
f_n	O(1)-Pr-O(1)	68.80	70.06
f_q	Cu-Pr-Cu	73.74	73.83
f_r	O(1)-Ba-O(1)	89.06	88.58
f_s	Cu-O(1)-Cu	165.26	161.9
f_t	O(1)-Ba-O(1)	56.01	55.24
f_v	O(2)-Ba-O(2)	88.37	88.29
f_α	Cu-Ba/Ho-O(2)	70.62	70.73
f_β	O(1)-Ba/Ho-O(1)	58.51	57.54
f_r	O(2)-Ba/Ho-O(2)	88.31	88.36
f_s	Cu-Ba/Ho-Cu	71.64	71.39

The lowest infrared phonon frequency at 138cm^{-1} and 168cm^{-1} are in the B_{1u} symmetry is due to the vibration of Pr and Ho atoms. Here the observed experimental frequencies very well with the calculated frequency at 140 and 160 cm^{-1} respectively. The calculated infrared phonon frequency at 290cm^{-1} is due to the vibration of Ba atoms and it is due to the bending vibration of O(1)-Ba-O(2) atoms. And it agrees very well with the experimental frequency at 300cm^{-1}. The calculated infrared phonon frequency at 317 cm^{-1} is due to the vibration of Cu atoms. The observed frequency agrees with the calculated frequency at 320cm^{-1}. The frequency at 336cm^{-1} in this symmetry is due to the O(1) atoms and it is due to the bending vibration of O(1)-Cu-O(2) atoms and it agrees very well with the calculated frequency at 340 cm^{-1}. The highest calculated infrared phonon frequency at 482 cm^{-1} in this symmetry is due to the vibration of O_2 atom and it is due to the stretched vibration of O_2 atoms and it agrees very well with the experimental values at 480cm^{-1}.

The calculated infrared phonon frequency in B_{2u} symmetry at 135 cm^{-1} and 300 cm^{-1} are due to the vibration of Pr and Cu atoms. These are due to the stretched vibration of PrO(1) and Cu-O(1) atoms and this experimental frequencies agrees very well with the calculated frequencies at 160cm^{-1} and 260cm^{-1} respectively. The calculated infrared phonon frequency at 158 cm^{-1} and 320 cm^{-1} are due to the vibration of Ho and O(1) atoms respectively. These vibrations are due to the bending vibration of O(1)-Ba –O(1) atoms and O(1)-Cu-O(2) atoms respectively and their experimental frequencies at 180cm^{-1} and 280cm^{-1} agrees very well with the calculated frequencies. The calculated infra-red phonon frequency in this symmetry at 198 cm^{-1} is due to the vibration of Ba atoms and Cu atom is 180° out of phase to Ba-O(1) atoms and this experimental frequency at 200cm^{-1} agrees very well with the calculated frequency. The highest frequency in this symmetry is 580cm^{-1} is due to the vibration of O(3) atom and it is due to the stretched vibration of Cu-O(3) atoms.

Table 4.4: Calculated Phonon Frequencies of Pr_x H_{01-x} Ba_2 Cu_3 O_7 (Values in the parentheses are experiment frequencies)

Symmetry Species	Frequency Using Normal Coordinate Analysis	Potential Energy Distribution (%)
1	2	3
A_{1g}(Raman)	290 (360)	f_c (58) f_d(24) f_a (11)
	330 (340)	f_c (55) f_l (30)
	355 (360)	f_l(71) f_d (12) f_p(11)
	400 (480)	f_a (51) f_α(20)
	502 (500)	f_a (68) f_u(20) f_c (10)
B_{2g}	300 (320)	f_a (56) f_k (14) f_a (10)
	418 (420)	f_a(52) f_k (20) f_β (6)
	330 (440)	f_c (60) f_e (14) f_β (6)
	450 (480)	f_m (70) f_e(18) f_β (10)
	530 (520)	f_m (70) f_l (21)
B_{3g}	298 (300)	f_l(45) f_n(31) f_c(20)
	415 (420)	f_a (42) f_g (25) f_n(15)
	456 (460)	f_α (68) f_n(21)
	500 (500)	f_n (70) f_a(24)
	535 (540)	f_m(65) f_u(20) f_d(11)
B_{1u}	138 (140)	f_a(81) f_e(16)
	168 (160)	f_p(54)f_n(30)
	290 (300)	f_p(62)f_β (18)f_n(10)
	317 (320)	f_a(40)f_g(30)f_n(10)
	336 (340)	f_β (40)f_p(30)f_m(26)
	482 (480)	f_a (60)f_e(21)f_m(16)
B_{2u}	135 (160)	f_a(71) f_e(20)
	158 (180)	f_β (46) f_d(18) f_b(10)
	198 (200)	f_α (52) f_a(17) f_e(24)
	300 (260)	f_a(55) f_d(18) f_n(10)

1	2	3
	320 (280)	$f_c(61)$ $f_e(16)$ $f_u(14)$
	400 (360)	$f_a(55)$ $f_e(14)$ $f_s(20)$
	580 (560)	$f_q(64)$ $f_r(16)$
B_{3u}	180 (140)	f_β (66) $f_p(25)$
	255 (260)	$f_n(65)$ $f_\alpha(21)$
	310 (300)	f_δ (68) $f_a(25)$
	330 (340)	f_δ (61) f_β (20) $f_m(10)$
	560 (560)	f_β (56) $f_k(17)$ $f_g(16)$
	581 (580)	f_k (80) $f_l(14)$)

The calculated infrared phonon frequency at 180 cm^{-1} and 255cm^{-1} in B_{3u} symmetry are due to the Pr and Ho atom. These vibrations are due to the stretched vibrations of Pr – O(1) and Ho – O(1) respectively. Hence the experimental frequency at 140 cm^{-1} and 260 cm^{-1} agree very well with the calculated frequencies. The calculated Raman phonon frequency at 310cm^{-1} is due to the vibration of Ba atom.

The calculated infrared phonon frequency at 330 cm^{-1} is due to the vibration Cu atom and it is due to the bending vibration of Cu(1)-Ba-Cu(2) atoms. The experimental frequency at 340cm^{-1} agrees very well with calculated frequency. The calculated infrared phonon frequency at 560 cm^{-1} is due to the vibration of O(1) atoms and O(2) atom is 180^0 out of phase to Ba–O(1) atom. And experimental frequency at 560cm^{-1} in calculated phonon frequency in this symmetry is 581cm^{-1} is due to vibration of O(2) atom and it is due to the stretched vibration of Cu-O(2) atoms. The observed value at 580 cm^{-1} agrees very well with the calculated frequency.

CONCLUSION

In this work the normal coordinate analysis technique have been adopted to give the evidence for electron, phonon interaction in the $La_2 Cu O_4$ and $Pr H_{01-x} Ba_2 Cu_3 O_7$. It is observed from the tables that the agreement between the calculated and observed frequencies were in good agreement with the systems that are considered here. This fact supports that the present vibrational assignments made for the infra red and Raman spectra are adequate. Therefore it is concluded that the normal coordinate analysis is the optically active vibrations of the vibrational spectra in cupurate oxides.

Lastly, these calculations yielded not only the zone centre phonon modes but also the stable dispersions. Hence, it also supports the strong electron phonon interaction in high temperature super conductor (HTSC). The vibrational frequencies calculated by the method of normal coordinate analysis are compared theoretically and experimentally, and they appears to be in good agreement that are further confirmed by the potential energy distribution calculation.

REFERENCES

1. J.G. Bednorz and K. A. Muller. *Z. Phys.* B 64(1986)189
2. M.K. Wu, J.R. Ashbirn, C.J. Torng, P. H Hor, R.L. Meng, L. Gao, Z. J. Huang, Y.Q. **Wangand** C.W. Chu, *Phys. Rev. Lett.* 58(1987) 908.
3. Z.Z. Sheng, A. M. Hermann, A. Elali, C. Almasan, J. Estrada, T. Datta and R. J. Matson, *Phys. Rev Lett.* 60(1988)937.
4. A. P. Ramirez, B. Batlogg, G. Aeppli R, J. Cava, E. Rietman, A. Goldman and G. Shirane, *Phys. Rev* B35(1987)8833.
5. B. Renker, F. Gompf, E. Gering, N. Nucker, D. Ewert, W. Reichardt and H. Rietchel, *Z. Phys.* B67(1987)15.

6. B. Renker, F. Gompf, E. Gering, G. Roth, D. Ewert, and W. Reichardt, Z. Phys. B 71(1988)437.

7. P. Bruesh and W. Buhrer, *Z. Phys.* B 70(1988) 70.

8. P. Bruesh and W. Buhrer, P. Unternaehre and A. Taylor, *Physica* C 153-155(1988) 300.

9. A. V. Belushkin, E. A. Goremychkin, Inatkaniec, I. L. Sashin, W. Zajac, A. R. Kadyrbaev and B. P. Michailov, *Physica* C 156(1988) 259.

10. S. Sugai, M. Sato. S. Hosoya, S. Uchida, H. Takagi, K. Kitazawa and S. Tanaka, *Jpn. J. Appl. Phys.* Part 2, Suppl. 26-3(1987)1003.

11. K. Ohbayashi, N. Ogita, M. Udagawa, Y. Aoki, Y. Maeno and P. Fujita, *Jpn. J. Appl. Phys.* Part 2, 25(1987) L423.

12. Torben Burn, M. Grimsditch, K. E. Gray, R. Bhadra, V. Maroni and C. K. Loong, *Phys. Rev* B35(1987)8837.

13. S. Sugai, M. Sato, S. Hosoya, *Jpn. J. Appl. Phys.* Part 2, 26 (1987)L495.

14. S. Blumenroder, E. Zirngiebl, J. D. Thompson, P. Killough, J.L. Smith and Z. Fisk, *Phys. Rev* B35(1987)8840.

15. M. Copic, D. Mihailovic, M. Zgonit, M. Prester, K. Biljakovic, B. Oreal and N. Brnicevic, *Solid State Commun.* 64 (1987)297.

16. B. Batlogg, G. Kourouklis, W. Weber, R. J. Cava, A. Jayaraman, A. E. White, K. T. Short, L. W. Rupp and E. A. Rietmann. *Phys. Rev. Lett.* 59(1987)912.

17. G. A. Kourouklis, A. Jayaraman, W. Weber, J. P Remeika, G. P. Espinosa, A. S. Cooper and R. G. Maines Sr., *Phys. Rev* B36(1987)7218.

18. K. B. Lyons, P. A. Fleury, J. P Remeika, A. S. Cooper and T. J. Negeren, *Phys. Rev* B 37 (1988)2353.

19. S. Zhang, He-Tian Zhou, Chongde Wei and Qin Lin, *Solid State Commun.* 66 (1988) 1085.

20. A. I Maksimov, O. V. Misochko, I. T. Tartakovosky, V. V. Timofeev, J. P. Remeika, A. S. Cooper and Z. Fisk, *Solid State Commun.* 66 (1988) 1077.

21. J. D. Jorgensen, H. B. Schuttler, D. G. Hinkds, D. W. Capone II, K. Zhang, M. B. Brodsky and D. J. Scalapino. *Phys. Rev. Lett.* 58(1987) 1024.

22. D. C. Johnston, J. P. Stokes, D. P. Goshorn and J. T. Lewandowski, *Phys. Rev.* B36 (1987) 4007 .

23. D. L. Rousseau, B. P. Baumann and S.P.S. Porto, *J. Raman Spectr.* 10(1981) 253.

24. B. Renker, F. Compf, E. Gering *et al., Z. Phys.* B 67, 15(1987).

25. B.N. Goshitskit, C. ADavydob, M. G. Zemlyanov *et a,* Fiz. Met, Metalloved. *Phys. Met. Metallogr.* 67, 188(1987).

26. I. Natkanets, A. V. Belushkin, Ya. Mayer *et al.*, Pis'ma Zh. *Eksp. Teor. Fiz.* 48, 166 (1988) [JETP Lett. 48, 181 (1988)].

27. L. Pintschovius, N. Pyka, W. Reichardt *et al., Progress in HTSC* 21, 36(1989).

28. F. Gervais, P. Echegut, J. M. Bassat, and P. Odier, *Phys. Rev.* B 37, 9364 (1988) .

29. Yu . S. Ponosov and G. A. Bolotin, *Pis'ma Zh. Eksp. Teor. Fiz.* 49, 16 [*JETP Lett.* 49, 16 (1988)].

30. M. Shimada, M. Shimizu, J. Tanaka *et al., Physica* C 193, 277(1992).

31. S. Ikeda, M. Misawa, S. Tomiyoshi *et al., Phys. Lett.* A 134, 191 (1988).

32. H. A. Mook, J.A. Harvey, and N.W. Hill, *Phys. Rev.* B 41, 764 (1990).

33. P.P. Parshin, M.G. Zemlyanov, I.E. Graboi, and A. P. Kaul', Sverkhpr. Fiz. Khim. Tekh [*Supercond. Phys. Chem. Tech.*] 2, 29 (1989).

34. P.P. Parshin, M. G. Zemlyanov, and P. I. Soldatov, *Zh. Eksp. Teor. Fiz.* 101, 750(1992) [JETP 74, 400(1992)].

35. C.T. Thomsen, M. Cardona, W. Kress, L. Genzel, M. Bauer, W. King and A. Wittlin, *Solid State Commun.* 64, 727 (1987).

36. H. Fuhrer, V. V. Kartha, K. G. Kidd, P. J. Krueger, and H. H. Mantasch, *Computer Programmes for infrared Spectrometry*, Vol. V, Normal Cordinate Analysis National Research Council of Canada, Ottawa 1976.

37. Jee Cs, Kebede, A., Nichols, D., Crow, J.E., Mihalisin, T., yer GH, Perez I, Salomon RE, Schlottmann P, *Solid State Commun,* 69:1139 (1988).

38. Fink, J., Nuckeer, N., Romberg, H., Alexander, M., Maple MB, Neumeieer JJ, *Allen JW, Phys.*, Rev. B. 42:4823, (1991).

39. Neumeier, J.J., Bjornholm, T., Maple, M.B., Rhyne JJ, *Gotass JA Physsssica* C1166:1191 (1990).

40. Friedl, B., Thomsen, C., Schonheeerr, E., Cardona M, *Solid Statte Commun.* 76:1107, (1990).

41. Friedl, B., Thomsen, C., Cardona, M., *Phys. Rev. Lett.,* 65:915, (1990).

42. Ltvinchuk, A.P., Thomsen, C., Cardona, M., *Solid State Commun,* 80:257, (1991).

43. Thomsen, C., Friedl, B., Cieplak, M., Cardona, M., *Solid State Commun,* 78: 727 (1991).

44. Zeyher, R., Zwicknagl, G., *Phys;* B 78:175 (1990).

45. Onari, S., Hidaka, S., Ara, T., Mori, T., *Solid State Commun.* 71:195(1989).

Bibliography

Banwell, C.N., *Fundamentals of Molecular Spectroscopy,* Tata McGraw-Hill Book company, New Delhi, Second Edition 1975.

Batlogg, B., G. Kourouklis, W. Weber, R. J. Cava, A. Jayaraman, A. E. White, K. T. Short, L. W. Rupp and E. A. Rietmann. *Phys. Rev. Lett.* 59 (1987) 912.

Bednorz, J. G., and K. A. Muller, *Z. Phys.* B 64, 189, 1986.

Belushkin, A. V., E. A. Goremychkin, Inatkaniec, I. L. Sashin, W. Zajac, A. R. Kadyrbaev and B. P. Michailov, *Physica* C 156 (1988) 259.

Bingal, W. A., *Theory of Molecular Spectra,* Wiley, 1969.

Blinc, R., and B. Zeks, *'Soft modes in ferroelectrics and anti ferroelectrics',* North Holland Publishing Co, Amsterdam (1970).

Blumenroder, S., E. Zirngiebl, J. D. Thompson, P. Killough, JL Smith and Z. Fisk, *Phys. Rev* B35 (1987) 8840.

Bruesh, P. and W. Buhrer, P. Unternaehre and A. Taylor, *Physica* C 153-155 (1988) 300.

Bruesh, P. and W. Buhrer, *Z. Phys.* B 70 (1988) 70.

Burn, Torben, M. Grimsditch, K. E. Gray, R. Bhadra, V. Maroni and C. K. Loong, *Phys. Rev* B35 (1987) 8837.

Chu, C. W., P. H. Hor, R. L. Mrg and L. Gao, Z. J. Huang and Y. Q. Wang, *Phy. Rev. Lett.* 58, 405, 1987.

Conley, R. T., *Infrared Spectroscopy,* Allyn & Bacon Ltd., 1966.

Copic, M., D. Mihailovic, M. Zgonit, M. Prester, K. Biljakovic, B. Oreal and N. Brnicevic, *Solid State Commun.* 64 (1987) 297.

Cotton, F. Albert, *Chemical Applications of Group Theory,* Wiley Eastern Ltd., Second Edition 1971.

Decius, J. C., *J. Chem. Phy.* 22, 1941 (1954).

Fink, J, Nuckeer N, Romberg H, Alexander M, Maple MB, Neumeieer JJ, *Allen JW, Phys. Rev.* B. 42: 4823, (1991).

Friedl, B, Thomsen C, Cardona M, *Phys. Rev. Lett.,* 65:915, (1990).

Friedl, B, Thomsen C, Schonheeerr E, Cardona M, *Solid Statte Commun.* 76: 1107, (1990).

Fuhrer, H., V. V. Kartha, K. G. Kidd, P. J. Krueger, and H. H. Mantasch, *Computer Programmes for infrared Spectrometry,* Vol. V, Normal Cordinate Analysis National Research Council of Canada, Ottawa 1976.

Ganguly, P., and C. N. R. Rao, *J. Solid state Chem.* 58, 193, 1984

Gervais, F., P. Echegut, J. M. Bassat, and P. Odier, *Phys. Rev.* B 37, 9364 (1988).

Gervais, Francois, *Physica. C.* 185-189, 2609 (1991).

Goshitskit, B.N., C.A. Davydob, M.G. Zemlyanov *et al.,* Fiz. Met, Metalloved. *Phys. Met. Metallogr.* 67: 188 (1987).

Gupta, S.L., V. Kumar, R.C. Sharma, *Elements of Spectroscopy,* Pragati Praksahan, India, 1983.

Herzberg, G., *Molecular Spectra and Molecular structure : Vol. 2, Infra-red and Raman Spectra of poly atomic molecules,* Van Nostrand, USA, 1945.

Horak M., and A. Vitek, *Interpretation and Processing of Vibration Spectra,* Willey Inter-science (1978).

Ikeda, S., M. Misawa, S. Tomiyoshi *et al.*, *Phys. Lett.* A 134: 191 (1988).

Jee Cs, Kebede A., Nichols, D., Crow, J.E., Mihalisin, T., Yer, G.H., Perez, I., Salomon, R.E., Schlottmann, P., *Solid State Commun*, 69: 1139 (1988).

Johnston, D.C., J.P. Stokes, D.P. Goshorn and J.T. Lewandowski, *Phys. Rev* B36 (1987) 4007.

Jorgensen, J.D., H.B. Schuttler, D.G. Hinkds, D.W. Capone II, K. Zhang, M.B. Brodsky and D.J. Scalapino. *Phys. Rev. Lett.* 58 (1987) 1024.

Khurana, A., *Phys. Today* 41 (No. 4), 21, 1988.

Kittle, C., *DIntroduction to Solid State PhysicsD,* John Wiley & Sons 1976.

Kleinman, D. A., *Phys. Rev.*, 118 (1960).

Kourouklis, G. A., A. Jayaraman, W. Weber, J. P Remeika, G. P. Espinosa, A. S. Cooper and R. G. Maines Sr., *Phys. Rev.*, B36 (1987) 7218.

Krantz, M., H. J. Rosen, R. M. Macfarlane and V. Y. Lee *Phys. Rev.,* 38: 11962, 11965, 1988.

Krishna Rao, K. V., *Physics of Solid State,* S. Balakrishna, M. krishnamurthi and R. R. Rao (Eds), Academic Press, London (1960).

Kulkarni A.D., and W. de wette, J. Prade and U. Schroder W. Kress, *Physica Review,* B5451, 5457, 1990.

Li, R., R. Ferile, G. Jakob, Th. Hahn and H. Adrian, *Phys. Rev. Lett.* 70: 3804 (1993).

Litvinchuk, A.P., C. Thomson and C. Cardona *Solid state communications* 83: 1992, 343-347.

Ltvinchuk, A.P., Thomsen, C., Cardona, M., *Solid State Commun*, 80: 257 (1991).

Lyons, K.B., P.A. Fleury, J.P. Remeika, A.S. Cooper and T.J. Negeren, *Phys. Rev.*, B37 (1988) 2353.

Maeda, H., Y. Tanaka, M. Fukutomi and T. Asano, *Jpn. J. Appli. Phys.* 27, L209, 1988.

Maksimov, A. I., O. V. Misochko, I. T. Tartakovosky, V. V. Timofeev, J. P. Remeika, A. S. Cooper and Z. Fisk, *Solid State Commun.* 66 (1988) 1077.

Marino, Y. and K. Kuchitsu, *J Che. Phy.* 20: 1809 (1952).

Michael, C., M. Herview, M. M. Borel, A. Gradin, F. Deslanders, J. Provost and B. Raveau, *Z Phys.* B68, 421, 1987.

Mohan, S. and A. Sudha, *Pramana J. Phys.* 37: 4, 327, (1991).

Mook, H.A., J.A. Harvey, and N.W. Hill, *Phys. Rev.* B 41: 764 (1990).

Nakamotao, K., *Infrared Spectra of Inorganic and Coordination Compounds*, John Wiley, New York, 1st edition, 1963.

Natkanets, I., A. V. Belushkin, Ya. Mayer *et al,* Pis'ma Zh. *Eksp. Teor. Fiz.* 48: 166 (1988) [*JETP Lett.* 48, 181 (1988)].

Neumeier, J.J., Bjornholm, T., Maple, M.B., Rhyne, J.J., *Gotass JA Physsssica,* C1166: 1191 (1990).

Ohbayashi, K., N. Ogita, M. Udagawa, Y. Aoki, Y. Maeno and P. Fujita, *Jpn. J. Appl. Phys.* Part 2: 25 (1987) L423.

Onari S, Hidaka S, Ara T, Mori T, *Solid State Commun.* 71: 195 (1989).

Parshin, P.P., M.G. Zemlyanov, I.E. Graboi, and A.P. Kaul, Sverkhpr. Fiz. Khim. Tekh [*Supercond. Phys. Chem. Tech.* [2, 29 (1989).

Parshin, P.P., M.G. Zemlyanov, and P.I. Soldatov, *Zh. Eksp. Teor. Fiz.* 101: 750 (1992) [JETP 74, 400 (1992)].

Pintschovius, L., N. Pyka, W. Reichardt *et al.*, *Progress in HTSC* 21, 36 (1989).

Ponosov, Yu . S., and G. A. Bolotin, Pis'ma Zh. Eksp. Teor. Fiz. 49: 16 [JETP Lett. 49, 16 (1988)].

Puranik, P. G., *Group Theory applications to molecular vibrations,* S. Chand & Company Ltd., First Edition (1979).

Putlin, S.N., E. Vantipov, O. Chmaissen and M. Marezio, *Nature,* 362: 226, 1993.

Ramirez, A. P., B. Batlogg, G. Aeppli R, J. Cava, E. Rietman, A. Goldman and G. Shirane, *Phys. Rev* B35 (1987) 8833.

Renker, B., F. Compf, E. Gering *et al, Z. Phys.* B 67, 15 (1987).

Renker, B., F. Gompf, E. Gering, G. Roth, D. Ewert, and W. Reichardt, *Z. Phys.* B 71 (1988) 437.

Renker, B., F. Gompf, E. Gering, N. Nucker, D. Ewert, W. Reichardt and H. Rietchel, *Z. Phys.* B67 (1987) 15.

Rose, S. D., *Inorganic Infrared and Ramnd Spectra,* McGraw-Hill Book Company, England, 1972.

Rose-Innes, A. C., and E. H. Phoderick, D *Introduction to SuperconductivityD,* Pergamon Press Ltd., 1978.

Rosseinksy, M. J., A. P. Ramirez, S. H. Glarum, R. C. Haddon, A. F. Hebard, T. T. M. Palstra, A. R. Kortan, S. M. Zahurak, A. V. Makhija, *Phys. Rev. Lett.* 66: 2830, 1991.

Rousseau, D. L., B. P. Baumann and S. P. S. Porto, *J. Raman Spectr.* 10 (1981) 253.

Schonland, D., D Van Nostrand, *Molecular Symmetry,* London, McGraw-Hill Book Company, 1st Ed (1965).

Sheng, Z. Z., A. M. Hermann, A. E. Ali, C. Almasan, J. Estrada, T. Datta and R. J. Matson, *Phys. Rev Lett.* 60 (1988) 937.

Sheng, Z. Z., A. M. Hermann, A. E. Ali, C. Almasan, J. Estrada, D. Datta and R. J. Matson, *Phys. Rev. Lett.* 60: 937, 1988.

Shimada, M., M. Shimizu, J. Tanaka *et al*, *Physica* C 193, 277 (1992).

Shimanouchi, T., M. Tsuboi and T. Miyazawa, *J. Chem. Phys.* 35, 1597, (1961).

Subramanyam, S. V., and E. S. Rajagopal, *High Temperature Superconductors,* Wiley Eastern Ltd., 1989.

Sugai, S., M. Sato, S. Hosoya, *Jpn. J. Appl. Phys.* Part 2, 26 (1987) L495.

Sugai, S., M. Sato. S. Hosoya, S. Uchida, H. Takagi, K. Kitazawa and S. Tanaka, *Jpn. J. Appl. Phys.* Part 2, Suppl. 26-3 (1987) 1003.

Thomsen, C., Friedl, B., Cieplak, M., Cardona, M., *Solid State Commun,* 78: 727 (1991).

Thomsen, C. T., M. Cardona, W. Kress, L. Genzel, M. Bauer, W. King and A. Wittlin, *Solid State Commun.* 64: 727 (1987).

Tokura, Y., H. Takagi and S. Uchida, *Nature,* 337: 334, 1989.

Wiberley, S. E., N. B Colthup and L. H. Daly, *Introduction to infra-red and Raman Spectroscopy,* Academic Press (New York), 1st edition 1964.

Wilson Jr., E. Bright, J. C. Decius and P. C. Cross, *Molecular Vibration,* McGraw-Hill Book Company, New York, First Edition (1955).

Wilson, E.B., *J. Chem. Phys.* 7, 1047 (1939).

Wilson, E. Bright, *J. Chem. Phy.* 9 76 (1941).

Wu, M.K., J.R. Ashbirn, C.J. Torng, P.H. Hor, R.L. Meng, L. Gao, Z.J. Huang, Y. Q. Wangand C. W. Chu, *Phys. Rev. Lett.* 58 (1987) 908, 1987.

Wu, T.Y., *Vibration spectra and structure of polyatomic molecules,* National University of Peking, Kun-ming China, 1939.

Zeyher, R, Zwicknagl G, *Phys*; B 78:175 (1990).

Zhang, S., He-Tian Zhou, Chongde Wei and Qin Lin, *Solid State Commun.* 66 (1988) 1085.

Index